# HOW TO WIN AND MANAGE
# CONSTRUCTION PROJECTS

# HOW TO WIN AND MANAGE
# CONSTRUCTION PROJECTS

A.L.M. AMEER MRICS, AAIQS, ACIArb
Chartered Quantity Surveyor & Arbitrator

*AuthorHouse™*
*1663 Liberty Drive*
*Bloomington, IN 47403*
*www.authorhouse.com*
*Phone: 1-800-839-8640*


© *2013 by A.L.M. Ameer MRICS, AAIQS, ACIArb. All rights reserved.*

*No part of this book may be reproduced, stored in a retrieval system, or transmitted by any means without the written permission of the author.*

*Published by AuthorHouse    03/04/2013*

*ISBN: 978-1-4817-8524-2 (sc)*
*ISBN: 978-1-4817-8523-5 (hc)*
*ISBN: 978-1-4817-8525-9 (e)*

*Any people depicted in stock imagery provided by Thinkstock are models, and such images are being used for illustrative purposes only.*
*Certain stock imagery © Thinkstock.*

*This book is printed on acid-free paper.*

*Because of the dynamic nature of the Internet, any web addresses or links contained in this book may have changed since publication and may no longer be valid. The views expressed in this work are solely those of the author and do not necessarily reflect the views of the publisher, and the publisher hereby disclaims any responsibility for them.*

# Contents

PREFACE ..................................................................................1
1. INTRODUCTION ..................................................................3
2. TYPES AND CONDITIONS OF CONTRACTS ....................5
3. RISKS IN CONSTRUCTION ...............................................11
4. ESTIMATION & TENDERING OF CONSTRUCTION PROJECTS ..........................................................................20
5. ADJUDICATION OF TENDER AND POST-TENDER NEGOTIATIONS .................................................................24
6. PLANNING, RECORD KEEPING, CORRECTIVE ACTIONS AND RECONCILLIATION ..............................27
7. PRICING OF ADDITIONAL WORKS/VARIATIONS .........31
8. CONSTRUCTION FRAUDS ................................................56
9. CONSTRUCTION CLAIMS, PREPARATION AND NEGOTIATIONS .................................................................76
10. CONSTRUCTION DISPUTES, MECHANISMS FOR RESOLUTIONS ................................................................104
11. CONCLUSION ...................................................................109

Profile of the Author ..................................................................111

# PREFACE

I find there is a necessity for me to write a Book to share with others my long and mixed experience of over 40 years in the construction industry with Employer, Consulting, Contractor and Teaching environments. As we move into 21st century, where certain studies indicate that in a couple of decades, for the first time in the history of mankind more than 50% of the population shall live in cities and this will create a new dimension in the scope of construction industry. It is also evident that as the credit crunch takes hold worldwide the concept of "Value for money" meaning the best possible value for money spent is being heard louder and louder and new forms of Disputes Resolution Mechanisms are being formulated in the construction industry. Needless to say that there will be substantial potential for growth for construction professionals.

The book shall be of immense use for young Engineers who may wish or decide to have an interesting and rewarding career as Cost Engineer in the Construction Industry. This book can also be useful for Engineers who seek careers in Construction Auditing/Alternate Dispute Resolutions. These two branches of Construction industry have seen a great growth in the recent past.

The book is mostly directed at Engineering professionals in the construction industry, while it will also be useful to have insight of the construction industry for small and medium size Contractors and individual clients who intend to build up their own villas. I also feel that it would be of immense help to young Engineers who seek careers in the construction industry, particularly in countries such as China, India, Middle East where Cost Engineering has not developed like say countries such as United Kingdom, Australia and Canada where many university offer degree Ccourses in Quantity Surveying and related fields.

There is another professional issue in publishing this book. I have to make explanation/substantiations for my statements regarding the two issues given below.

a) Once I made a presentation on the topic "Construction as an option to international Investors (the summarized version was published in Building Economics-AIQS Dec.2011) and I stated that in such cases the Quantity Surveyors can Work as advisors to them and earn in excess of US $ 100,000/= a week. Some Quantity Surveyors were sceptical or couldn't comprehend and it became joke to them. My chapter on Tender Adjudication and Post-Tender negotiations answers this aspect.

b) I made another point that if someone could do Technical Auditing of a big project that would be enough for his entire life and quoted a Lawyer in Sri Lanka in th 1960's who did only one case in his whole career. He earned so much that he never sited the Courts in rest of his life. Again, some people were sceptical, couldn't comprehend this. My case studies given under Construction Frauds shall give an indication regarding the possibilities of the amount of savings in Technical Auditing (if one is very smart enough) and going by the Industry standards (1/3 to Auditor & 2/3 to Client) shall clearly give an answer.

# 1

# INTRODUCTION

Construction which originated from the requirement of a shelter which is one of the basic needs of the human race and is therefore is as old as the human civilization itself. Over the years, construction has evolved into palaces, townships, capitals and we are in a very advanced stage where. Construction is big business and forms a substantial chunk of the labour force and the GDP of every nation. Certain studies indicate that in a couple decades as a first time in the planet more than 50% of the population shall live in cities. This creates a situation, where village people shifting to towns in large numbers. This has already commenced in China to a very great extent. This newly settled or to be settled people require housing and other infrastructure which the Governments compelled to make. Construction procurements have evolved into new types of Contracts involving Government and private partnerships. Private investors with Billions of Dollars are slowly turning to construction owing to partly of the turbulent nature of the financial markets in general and partnership arrangements. Value Engineering is also taking ground. The voice of "Value for money" or the requirement of best possible value for the money spent is being heard louder and louder. Cost of financing construction which is essentially done by the banks is becoming more and more expensive.

There is what famously known as the triple "C" connection in construction. The interaction between the three Cs namely Client, Consultant and Contractor shall determine the success or failure of Construction projects.

In spite of many Construction related Professional Institutions, construction companies with the army of learned construction professionals, the industry as a whole is full of problems. So much so, that the litigation in construction is only comparable to that of Extra-marital cases. If not for the problems that are infested with the construction industry, it would have become the best option for International investors with billions of dollars, particularly in to-day's economic climate where the traditional options such as sovereign bonds and stocks are in a state of turmoil to say the least. The industry as whole is surviving because of shear necessity of infrastructure for the ever growing population.

The book intends to give some fundamental aspects of construction in a more practical sense in the modern world which can be useful both to the lay man who intends to build up a villa for himself to that of construction professionals.

# 2
# TYPES AND CONDITIONS OF CONTRACTS

The Construction professionals should be aware of the Conditions, types of Contracts in order to advice the suitable procurement route for the Employers. The key issues that govern the construction are Cost, Time and Quality. It is often difficult to satisfy all three of these with one option, therefore consideration to be given to the Employer list of priorities and the implication of cost, time, quality to get the best route for the Employer.

## Conditions of Contract

Construction projects create particular challenges and risks for both the Employer and the Contractors. Large sums of money must be committed over time frames that can be years. There are many variables that can be impossible to accurately predict in advance and this can have a big impact on the ultimate cost and time of a project. Further the technical designs can be complicated and require careful co-ordination between several specialist sub contractors and design consultants both before and during the execution of a project.

Therefore, it is no surprise that the construction industry is full of disputes and insolvencies. In order to manage and mitigate these risks, the use of standardized construction contracts has become common place. Standard form of construction contracts attempt to impose certainty regarding the consequences of future events and to appropriately allocate risks as between the employer and contractor. There are various types of conditions of contract. Each country has its own. Many countries and cooperates have tailored their conditions of contract based on FIDIC with changes to suite their particular needs.

FIDIC Conditions of Contracts are widely used in Construction projects. These conditions of Contract have undergone changes for the best part of 60 years. In addition to its widespread familiarity within the international construction, the key advantage of the FIDIC range of contracts has been the generally accepted fairness of allocation of risks between the employer and contractor. The primary concept has been that any particular risk should be borne by the party best suited to manage same,-the contractor to take the construction risks which he can reasonably estimate, while the employer takes the risks of the unforeseen circumstances which cannot be accurately estimated in advance. This approach means that the employer only pays for those circumstances which actually arise, rather than potentially paying large sums for the contractor to cover all potential risks by adverse pricing. However, there is one exception, where FIDIC broke this tradition in 1999 by publishing the Conditions of Contract for EPC Turnkey Projects commonly known as "the Silver book". This Contract expressly passes additional risks to the contractor in order to maintain certainty of price, time and performance for the employer. FIDIC's justification for moving away from its traditional approach was to mention that many private clients wanted greater certainty of the final price and the time for completion and were willing to pay a premium to contractors in return. FIDIC insists that they are responding to market demand for such kind of commercial arrangement.

The new models introduced in 21$^{st}$ century are as follows:

- FIDIC Conditions of Contract for Construction(New Red Book)
- FIDIC Conditions of Contract for Plant and Design/Build (New Yellow book)
- FIDIC Conditions of Contract for EPC Turnkey Projects(Silver book)

- FIDIC Short Form of Contract (Green Book)
- FIDIC Condition of Contract for Design, Build and Operate Projects ( Gold Book)

## Types of Contracts

**1. Re-measure Contracts**

These are projects where the Employer produces usually with their Consultants the Design and tender documents. This implies that the design is complete. There can be two types of Contracts under this category

    1.1. Re-measure with fixed unit Rates

The project is usually tendered with approximate quantities and upon completion the quantities are re-measured and the final contract price is achieved. The unit Rates are fixed during the Contract period. This is mostly preferred by the Govt. departments owing to transparency, fairness ect. This actually gives a level playing field for all contractors and the construction risks are largely reduced.

    1.2. Re-measure with fixed unit Rates with escalation Clauses

These are essentially the same as above, but there is provision for the escalation of prices during the Contract period and the same is added to arrive the Final Contract price.

**2. Lump sum Contracts**

The Design and Tender documents are prepared by the Employer, usually through his Engineers and Contractors are asked to quote a Lump Sum value for the project. The Tender documents usually include drawings, specifications, Bills of Quantities and Conditions of Contract.

## 3. Design & Build Contract-Turn Key Project

The project brief and requirements are clearly defined. The Contractor is responsible for both Design and execution. A word of caution is required here, as in some Contracts particularly under FIDIC Silver Book; all risks including underground and other unforeseen conditions are transferred to the Contractor to maintain certainty of time and price.

## 4. Cost plus Contracts

The entire cost of the project is reimbursed to the Contractor with an agreed percentage addition for his overheads, attendance and profit. The advantage of this kind of arrangement is that design of the project need not be complete and early start can be made. The disadvantages are that there is no certainty of the Final price and there is no incentive for the Contractor to reduce the cost and a situation where the Contractor gains when the cost is increased is created. There is also a possibility for idle labour or under performance by labour and equipment at site. These types of projects are now out dated as Employers are asking the best possible value for money. These types of Contracts are still available, particularly in the United States of America.

## 5 Cost plus fixed fee Contracts

Exactly similar to cost plus but a fixed fee is paid to the Contractor instead of a Percentage addition.

## 6. Maximum guaranteed price.

Design brief and other details are submitted along with a maximum guaranteed price for the work. That is the limit to the Final Contract price. If the Final price is less than that then a saving arrangement is pre agreed with the Employer and the Contractor.

## 7. Build, own, operate, transfer (boot)

These are mega projects with the requirement of heavy investments such as harbour, factories or major high ways etc. Usually the investor builds this project, own, operate for an agreed period and then hand over the project. This is comparatively new and many investors are interested in such projects. This is usually a Govt. private partnership ventures. This type of arrangements are becoming popular particularly, in developing countries as the advantage is investors are bring in capital, which provides employment for locals both during construction and in operation and finally the entire project is handed over free to the Client.

## 8. Management Contracting

In this kind of arrangement, usually there will be a main Contractor and all works shall be split into work packages with realistic provisional sums. The packages shall be awarded to other Contractors under the main Contractor. The main Contractor receives an agreed percentage or amount for his, overhead, profit and the facilities he would be expected to fulfil during the project.

## 9. Contract Management Contracts

The scope can be defined to a very good extent but gives room for changes by the Employer during execution. The Employer takes full responsibility usually thorough a Project management Consultant. Work is split on trade lines and given as separate packages. Each package forms a sub-Contract. The fee arrangements are agreed initially. This is essentially a cost plus type of Contract.

## 10. Partnering Contracts

The Client assumes the overall management of the Contract. The design and scope is not necessarily complete and project start with an open book with a proposed budget, Time and Fee arrangements for all participants. Usually there will a management Contractor or one or two main Contractors with their own work packages and other work packages. Each work package shall form a Contract or may be treated as a sub-Contract under the existing main Contractors. This is usually cost plus type of Contract. Incentive schemes also can be there to create value for money.

It should be noted that there are advantages and disadvantages in each type of Conditions of Contract and the type of Contract. The Procurement arrangement of one project is not necessarily the same for another. The Quantity Surveyor should be in a position to advice their employers regarding the suitable method of Procurement taking into consideration of their individual needs and concerns.

There are new Employers such as International Investors coming to construction owing to the drying up or much higher Risks in their traditional options. These people have Billions to invest and traditional Risk takers for higher gains. A completely different approach say modifications to Contractor payment terms, which may result in substantial discounts from the Contractor can be an attractive option for such people.

## Types of Measurements

The methods of measurement in the construction industry are as follows:

1. Standard method of measurement (S.M.M.) generally for buildings
2. Principles of Measurement (International) For Works of Construction.(POMI) for buildings and Civil Engineering Works
3. Civil Engineering Standard Method of Measurement (C.E.S.M.M.) Generally for Civil Engineering projects.

# 3

# RISKS IN CONSTRUCTION

There are many Risks in construction, ignoring same shall be a financial disaster for Contractors. A brief summary along with ways and means of meeting same is given below

## 1. Risks when scope & Design not fully clear.

Knowing the full scope is a fundamental requirement to value the cost of a project. The situation, where the scope is not fully clear at the tendering stage is a present day reality in certain projects. This Risk has increased owing to the challenges that the Industry is facing in securing Contracts in an increasingly difficult environment.

Construction companies are also tempted, owing to the value and the competitive environment that they are facing and more and more are prepared to take such risks and even contemplating to reduce their usual margins to keep the company afloat at the difficult times.

There is another element to these puzzle-new investors who are not that experienced in construction field, but are willing to invest vast sums owing to the drying up of their traditional options such as severing bonds and stock markets. They are bringing in another problem that they are unable or unwilling to wait till their professionals finalize the scope in all respects such as design, specifications etc. They want the construction to start at the earliest, which is jokingly called "what they are asking to-day is actually wanted yesterday".

The problem is further complicated by the presence of sub-standard professionals in the Employer and Contracting organisations as companies are finding it difficult to sustain top talent, again the result of financial constraints. The difficulty with this group is that they are more than willing and pushing their Employers to start the project at the earliest, as they see this as their best bet for their own survival.

*This Risk can be overcome as follows:*

a) *The Contractors should always ask and get the complete scope. This is a fundamental requirement to price a project correctly, particularly now as margins are becoming lower and lower.*
b) *In the event, they are unable to get the full scope the next step is to divide the project scope into two portions namely*
   b.1. *Portions that the scope and design are fully clear so the same can be estimated correctly.*
   b.2 *the other portion where the scope is not that clear to be allocated with realistic Provisional sums.*

*There is no necessity for the Contractor to take the risk of the unknowns as it is not their making in the first place. Further it is always advantages for the Contractor to have more and more amounts under Provisional sums as he gets his attendance and mark ups basically without much effort and in some cases without much financial expenditure.*

It is often said that Provisional sums is the prescription of a tender document meaning zero provisional sums is the best tender document. It is common knowledge in the construction circles that all additional works or works which are finalized after the award of Contract or during construction period is always inflated at least by 10%. If the Employer is not patient enough to wait for their Consultants to prepare the tender document properly, then it is fair for him to pay a premium for starting prematurely.

Again since the Employer can't wait for his Engineers to finalize the full scope and prepare the tender documents accordingly, which often takes for certain projects up to one year and willing to start half way(say with around 30% under provisional sums), then it is reasonable for him to expect the overall Contract value is in excess of say around 5%. This may be acceptable to some Employers as they are getting the product earlier than anticipated and early return on the capital can off-set such loss.

## 2. Risks regarding underground Conditions

There is a risk with regard to underground conditions. The Contractors are often given the soil investigation report of the plot with the tender documents. These Soil reports contain so many qualifications that the report is of no value in the Contractual sense.

There are costs such such soil improvements, different type of foundation than estimated with the available data etc.; this can have a huge impact particularly in Design & Build Contracts. If we consider a 10 Storey Hotel Construction project, the actual soil met with and the soil report initially given at the time of tender may be entirely different. The requirement now may be improvement of soil or other major changes to foundation than what was estimated at the time of tender. This cost can run into millions and what was estimated to be a project with a reasonable profit can end up with huge loss.

The Contractors should be aware of the recent FIDIC edition for Design & Build Projects (EPC Turnkey Projects) where, contrary to the construction norms, the entire risks pertaining to the underground conditions have been transferred to the Contractor. This creates a situation that the Clients end up paying such risks even the same is not there. FIDICs explanation was that they are reacting to the market situation, where Employers are asking for the certainty of time and price and hence let them pay a premium.

It is not my intention to argue the fairness or moral behind all these, my concern is that Contractors should be aware of such risks, do the necessary provisions in pricing the project.

Some conditions of Contract do have a remedy for this situation with Clauses under unforeseen circumstances or conditions in the Conditions of Contract which entitles the Contractor for additional cost & time in the event the Contractor encounters a situation which could not have been expected at the time of tender by an experienced Contractor. Here again, the onus of proof lies with the Contractor, which is practically difficult with the qualifications given with the soil report.

My personal opinion substantiated by certain court decisions is that the presence of the above Clause in the Conditions of Contract shall benefit the Contractor rather than the Employer if the same happens to be underground conditions and if it could be proved that the Contractor has no access to project site to verify the given soil data before the tender submission and the liability of the Contractor is to build for the intended purpose.

*It is the duty of the Contractor to check the Clauses, which deals with unforeseen underground conditions and price such risks. One way of mitigating this risk in Lump sum Contracts is the inclusion of realistic provisional quantities for Substructure and the same is subject to re-measure. As the substructure portion can be finalized much earlier in the execution stage, the final amount of the Contract can be adjusted at an early stage.*

## 3. Escalation of Materials, plant and labour prices

Construction of a medium sized project usually takes around one year. Lot of things can happen within one year, particularly at present. The prices are always on the increase. The only exception is the lowering of prices after a steep rise during the period of stabilization. This is one aspect which requires careful consideration, while pricing by the Contractor.

*It is the duty of the Contractor to have list of all suppliers and it is also important to cultivate good relationship with the suppliers. As explained before, it is of primary important to estimate correctly all quantities of materials and a material status chart to be prepared to know when these materials shall be required at site in order to prepare materials ordering schedule during construction. Materials purchase negotiations to be carried out on that basis. It is very important to have agreement with suppliers to have fixed price agreement throughout the Contract period. Payment terms to be based on Interim payment receipt. The ideal case would be no escalation of prices during the Contract period and suppliers' payments can be met with the receipt of Interim/monthly payments. This will reduce the financial difficulties such as the necessity of payment upfront for materials.*

*The escalation of prices for labor and plant can be controlled much easier than materials, particularly with In-house labor and plant. Careful consideration to be given in the event labour & plant are to be hired from outside parties as again their agreement needs to be made similar to that of materials as given above.*

*Escalation of prices during the Contract period is a reality particularly in Contracts with longer duration. The Contractors usually price this risk by assuming certain percentage increases across the board for materials. Escalation Clauses in the Conditions of Contract, where there are provision of payment for increases in the base tender price of materials during the Contract are now available in certain Contracts. Employers need to be convinced the benefit and fairness of such Clauses.*

## 4. Risks with Specification.

In the good olden days, reputed companies were supplying materials and the construction methods with such materials were known and pricing work was easy. Now lot of companies are producing materials and there is a wide price difference( even up to 100% in some cases) for some materials and plant meeting the same specification. Another new element has come into the scene-the price difference between equipment produced and assembled in the country of origin and the equipment assembled in another country. There is another factor in the equation, some Consultants insisting on specific products and not prepared to accept alternatives meeting the same specification. This can lead to a financial disaster to the Contractor, particularly Contracts with equipment, in the event Contract is priced based on one country product and the requirement at site is a product from another country. Both products have the same Technical specification.

*The Contractors have to read carefully the specification and has to indicate in the tender, which country products they have used in their tender price build up. This can allow the Contractor to claim in the event the Consultant insisted of another country product.*

## 5. Political Risks

### 5.1 Political stability

*The Contractor has to consider the political stability of a country before doing any construction. The political instability can create situations such as the very existence of the employees at site may not be possible, let alone project. Some Insurance companies cover for such risks with a very high premium and the same can be passed to Client as a cost element. However, it is not advisable to do any work in a country where there is no guarantee of political stability for five years.*

## 5.2 Constrains on the availability and employment of expatriate staff

*In certain countries there may be government pressure to employ locals at all levels. Careful consideration to be given regarding the availability of such resources in the host country, their skill levels, the average salaries etc. as this may involve additional costs and the same to be considered in the pricing.*

## 5.3. Insistence on the use of local firms and agents

*The Contractor has to analyse the cost variations in such an eventuality and include same in the pricing.*

## 5.4. Customs and import restrictions and procedures.

There may be restrictions on the import of products; the procedure for such exercises can be long and complicated in certain countries.

*The project requirement to be studied and the risk to be considered in the pricing.*

## 5.5. Difficulties in disposing of plant and equipment.

Some countries insist that all plant and equipment brought to the project to be disposed locally upon completion and not allowed to remove.

It is not always possible to obtain the calculated or reasonable second hand value in the local market for these equipment and plant and there is no way to take them back. So there is a loss—the short fall in the expected return of these equipment and plants.

*This is a cost element and the same to be considered in the pricing.*

## 6. Logistical Risks

Availability of resources-, materials, construction equipment, spare parts, fuel, transportation facilities.

*This is a cost element and the same to be considered in the pricing.*

## 7. Construction Risks

7.1 Uncertain productivity of resources
7.2 Weather and seasonal implications
7.3 Industrial relation problems

In certain countries the company norms in relation to labour and plant cannot be applicable to climatic conditions such lower temperature, lower visibility, too much rains. The host labour can be subject to local industrial laws which may not be the same as that of the company.

*The Contractors should be aware of the above facts, make necessary adjustments in the pricing.*

## 8. Financial Risks

8.1 Inflation
8.2 Availability and fluctuation in foreign exchange
8.3 Delay in payment
8.4 Repatriation of funds
8.5 Local taxes

There are possibilities for some or all of the above risks in certain Contracts.

*The Contractors should be aware of the above facts, make necessary adjustments in the pricing.*

## 9. Third Party Liabilities

The Contractors should be aware of third party liabilities that can come into play in certain projects.

*The Contractors should be aware of the above facts, make necessary Insurance cover that may be required and the same to be included in the costing of the project.*

# 4

# ESTIMATION & TENDERING OF CONSTRUCTION PROJECTS

Estimating the cost of a project is a highly professional job. A project can have many risks, some of which may come to light in a later stage and situations that the Contractor may not able to get compensation from the Client. Good faith is totally absent in most if not all standard Conditions of Contract. To quote a simple example, everybody knows that money is required to do anything and that includes construction. In most Conditions of Contracts, even non-payment by the Employer requires notice from the Contractor to slow down/stop work.

Let us take another example, a quotation to cut a tree in a house garden. It is common knowledge that every house has knifes, ladders, axe etc. The Contractor cannot say that in good faith assuming, every house has equipment that are necessary to do the job, I gave a low price. The view of the Conditions of Contract is the Contractor has to do the job by himself without any help from the Employer. The risk is totally to the Contractor.

Therefore, estimation is very important and hence good and efficient people must do the job. A company can become bankrupt owing to wrong estimation of a single project. Exgratia payments are only in text books and never seem to have been implemented, certainly not in govt. funded projects.

A Construction company big or small must have it norms for plant and labour. The same have to be revised on the basis of feed backs from sites and updated regularly. All plants to be listed along with their book values. Sub and Specialist Contractors, materials suppliers lists are to be updated. Market treads and new materials in the market along with their uses to be maintained.

Construction Risks are often overlooked but it can sometimes effect the entire operation of the company. Construction risks and ways and means of mitigating same are described in details under risks in construction.

The estimating team members should visit the intended construction site to study the site situation, access, soil conditions, and other factors that will affect the execution of the project. It is important for the estimating team to calculate the exact requirement of the project in terms of material labour and plant and submit the management a proposal as follows:

1. Name of Project
2. Employer
3. Consultant
4. Total materials cost
5. Total Labour cost
6. Total Plant cost
7. Possible risks and the cost effect
8. Nett project cost without Overhead and Profit
9. Base programme
10. Whether the project can be executed with in house resources if not the details requirement of hired labour and plant.
11. Cash flow
12. Tender Index based on the following:
    12.1. Completeness of the Design/scope
    12.2. Site visit and general position of site;
    12.3. Risks
    12.4. Urgency of this work
    12.5. Previous experience with this Employer if any.

It is the function of the top management to fix the Overhead and profit or any other discount that can be done for this tender.

Other Risks can be as follows:

Economic slow down
Increased competition owing to recession
Damage to reputation owing to badly managed projects in the recent past.
Failure to attract or retain good talent
Regulatory or Legislative charges
Third party liability
Injury or death to workers
Cash flow/Liquidity/capital availability/credit risks.

There are other Risks for which nobody is responsible such as Wars, earthquakes, Forcemajure, etc. These Risks are often taken care of in the Conditions of Contract

This does not necessary mean that every risk is ever present, which will be similar to cannot get out if one thinks of accidents on the roads. The concept is, these risks may be there and should be taken care of.

The duty of the Estimating team is to cost the entire project and breaking down to elements such as

Materials
Labour
Plant
Risks and their financial implications
Assumed overheads

Assumed Profit
Total Tender Value.

The estimating team is also to prepare detailed cash flow to give an indication of the financial needs in the event of implementation of the project.

# 5

# ADJUDICATION OF TENDER AND POST-TENDER NEGOTIATIONS

The top management personal shall do the adjudication before the formal submission of tender. The topic of discussions can be summarized as follows:

1. Status of the order book. Are we in real need or desperate to get this job. This can be owing to the situation of the company or the familiarity of the project (similar ones were successfully done previously, thus have sufficient experience)
2. Do we have the resources & Technical knowhow in house to do the job.
3. What would be the additional cost (as a percentage) which we have to encounter in the event of outside hiring of labour and plant.
4. How is the Employer?. Good pay master, some what ok., or plainly bad
5. Type of Conditions of Contract—Familiar, completely new
6. Risk factor.
7. Can we do the project with our regular suppliers giving the bulk of the materials or do we have to import or look for new suppliers.
8. Possibility of using materials already in our stores profitably
9. Proximity of the site. No need for labour accommodations.
10. Advance payment amount, Payment terms, the approximate financial charges based on initial Cash flow forecasts.

11. Major Plants. Has any reach Zero value if so the list of such plants in order to consider possible discounts on such plants.

After discussing the above points, the management can decide about the discounts if any or the adjustments to Overheads and Profit and make the submission of the tender

The main point in staying successful is to understand the full extent of the scope; the complexities, the risks and price all to avoid any difficulties while execution.

## Post-Tender Negotiations

This is phenomena particularly with individual Employers as every body wants to cut corners and interested in the best possible value for money.

The top management with sufficient authority to be in charge of such negotiations. In the post—tender negotiation, there is a possibility to know the near rivals tender amounts as tenders are already opened.

The top management during such negotiations can contemplate substantial reductions/discounts in addition to what is already given in the tender value on items such as

a) Overheads
b) Profit
c) Mark up on Provisional sums
d) Mark up on Nominated sub—Contractors

In the event the Conditions of Contract can be amended to:

1) Increase of Advance payment to say 15% of the Contract price rather than the 10% normal.

2) More favourable monthly payment terms—say to be paid within 30 days upon Engineer's certification.
3) In projects where the bulk of the materials/equipment to be imported overseas, payment for materials at site to be considered when the same is in shipment in high seas rather than delivered to site.
4) Higher Invoice value for materials at site. Say 90% instead of the usual 80%.
5) Deduction of the Advance Payment to be based on the value of work done rather than the elapsed Contract duration.

It is to be noted that the Employers may be willing to consider the above as they already have the budgeted amount.

The bottom line is, Construction projects are carried out with financial outlay upfront by the Contractors in order to get a reasonable profit on their investment. If the project can be financed in full by the Employer's money upfront then there is no investment from the Contractor and thereby his profit margin can be reduced as it amounts to net profit.

# 6

# PLANNING, RECORD KEEPING, CORRECTIVE ACTIONS AND RECONCILLIATION

## Planning

Any project which is not properly planned is certainly planned to fail. Therefore, planning is very important for any project.

### Base Programme

Construction planning is a fundamental activity in the management and execution of construction projects. It involves the choice of Technology, the definition of work tasks, the estimation of resources and durations of individual tasks, and the identification of any interaction of different work tasks, the extent to which sub—contractors will be used and the time that is likely to take for equipment/materials, which are off the self-items requiring Engineer approvals, ordering and delivery periods. The requirements of any owner supplied equipment/materials that have to be included in the overall plan. It is very important that the original programme should fulfil the criteria for reasonableness and completeness.

The Programme has to be upgraded on a regular basis. This is a requirement to track the progress of work, corrective action to be taken in the event of any problems which will have an effect on the final completion date. It also helps to monitor payments, upgrading of company norms based on actual production at site, documentation such as relevant studies regarding delays, if any, notices, follow up actions, updating of the programme for submissions of Interim Extension of time and associated costs. Monitoring the programmes also helps to make corrections regarding the short fall in the original allocations of resources.

It is better to arrange a kick off meeting initially, to set a design or set off a system of communication, roles and responsibilities of personals etc. along with the general administration mechanism of the Project. The site meetings, which are a regular feature, correspondence and filing, have to be organized in good manner.

**Record Keeping.**

A good site organization requires good record keeping for any medium or major project. The list of requirements is given below:

1. 1. Original Signed Contract documents including Contract drawings, Specifications, Bills of Quantities, Post-tender clarifications, Soil reports
2. Base Programme, Contract Milestone Schedule, and Detailed programme
3. Short term Schedules and updates.
4. Materials/Equipment delivery Charts particularly materials/equipment which are to be ordered overseas and have delivery periods Original Tender Estimate, Build ups
5. Issued for Construction drawings and all revisions
6. Shop drawings, originals, all revisions and re-submissions
7. Engineers Instructions, change orders, Estimates, approvals
8. Daily Time records for Labour and equipment
9. Progress payments
10. Forecast-to-complete estimates and updates
11. Site meeting records
12. Transmittals
13. Test reports of works and materials
14. Request for Information

15. Sub-Contractors quotes, contracts, purchase orders, correspondence
16. Productivity Reports/analysis
17. Accident reports
18. Quality reports
19. Daily production logs, weather reports
20. Progress site photographs.
21. Progress reports, weekly, monthly, quarterly
22. Notice of claims for delays/extra cost by the Contractor.
23. Follow up actions that are required as per Conditions of contract.

**The Contractor should have the following plans during the implementation**

**1) Communications Plan**

Create a document showing that needs to be kept informed about the project and how they will receive the information. The communication system for example any letter from Engineer to Contractor should deal with only one subject. All Site Instructions given by the Engineer's representative such as Inspectors to be confirmed by the Engineers Instruction. The same to be numbered. Any Instruction pertaining to additional work/Variation to be initially costed by the Engineer to see for example the approximate cost & time requirements and then to see whether the same could be accommodated within Budget and time before the consultation with the Employer regarding implementation. Only approved additional works/Variations to be as additional works/variation.

A Format for weekly meeting to be devised. The most common mechanism is a weekly or monthly progress report, describing how the project is performing, milestones achieved and work planned for the next period.

## 2) Monitoring Plan

The project has to be monitored both to see both its adherence to the original plan and cash flow forecasts. Any discrepancy to be analyzed to see whether the delays if any are due to the Contractor/ Employer.

## 3) Action Plan

There is no point in having a monitoring plan without follow up action. Follow up action to be followed as per Conditions of Contract at the earliest. The Feedback to be compared with the company regarding productivity of Labour and Plants and the same to be updated if there is such a requirement.

## 4) Reconciliation

Reconciliation of Materials should be adapted for all construction projects. That is materials delivered to site and their uses in order to identify the wastage factor and to see whether the same is within the original estimate. This can also create a system to eliminate theft or waste. Reconciliation also helps to find out about any abortive work at site. Reconciliation of labour and plant (Actual Vs Planned) shall give a clear, idea whether the updating of the norms is a requirement or under performancee is prevailing at site.

# 7

# PRICING OF ADDITIONAL WORKS/VARIATIONS

Experience has shown that Additional works and Variations are part and parcel of any type of Contract. All Construction professional should be aware of the existence of same and must have basic ideas of pricing same.

## GENERAL

Variations are ever present in any construction projects. There are many reasons for such occurrence. Each construction project is new and unique in its composition, such as construction, Contractor, Consultant and Employer. It is humanly impossible to study all the needs of the Employer and produce a Tender/Contract document for implementation.

As the construction by its nature deals with different characters and personalities, additions and omissions from the Contract scope bound to be present. These changes to the Contract in general termed as Variations as this is the result of variance.

## CONTRACTUAL POSITION ON VARIATIONS

As constructions are always complex, there is always provision in the Contract for the Engineer to change the quantity and quality of work. In the absence of such a provision, the Contract shall be rigid and each Variation shall require a new Contract, which is not a feasible option.

# MECHANISM OF PRICING VARIATION.

As Variations are inevitable, there is a necessity to provide a mechanism or guide lines to price such variations. This is an important requirement in the Conditions of Contract owing to:

The Variations had always been and continue to be part of any construction or maintenance contract.

It is common knowledge in the construction circles that Variations are always priced higher in comparison to tender prices—at least by 10% and it is often considered to be a playground for smart Contractors.

Pricing of variations normally proceed on the following ways

3.1. Priced based on Rates available in Contract, if applicable.
3.2. If no such Rates are available in Contract, then Contractor and the Engineer to agree a suitable Rate.
3.3. If such agreement cannot be reached, then the Engineer has the power to fix the Rate.

### *3.4 FINANCIAL EFFECT OF VARIATIONS*

**The Employer utilizes his resources and money for the Engineer, to prepare proper tender documentations, invitation of tenders, tender analysis report, Contract documentation etc. to get value for his money.**

**As mentioned above, variations often considered a playground for Contractors and a smart Contractor shall use same for his maximum advantage. Thus, the presence of excessive variations and a smart contractor shall invalidate the above concept of "value for money"**

The writer feels that it is very important that sufficient discussions had taken place with the Employer and the

Engineer and the scope of work is fully comprehended before the tender stage. The tender queries and clarifications along with the quantumn of Provisional sums shall give an indication of the standard of tender documentation.

The tenders should be fully analyzed before the award to keep variations down to an acceptable level and to have a proper pricing mechanism for such variations.

## *PROCEDURES IN THE EVENT OF VARIATIONS*

Guide lines for Variation procedures for Consultant/Employer for Re-measure types of Contracts during the post Contract stage.

1. **Anticipated Variations.**

When a Variation arises or considered at site, the Site Consultant Quantity Surveyor, without waiting for the Contractor to prepare the anticipated cost, should give a no. and prepare an approximate estimate for same and submit to the Engineer. with the following

- a) Brief description of Variation.
- b) Approximate cost
- c) Whether the amount can be met with anticipated savings in the project or new additional funds would be required.
- d) Time impact (whether the additional work can be carried out during the Contract period or additional time required)

While preparing the approximate estimate, the Consultant QS would know what are the additional information such as drawings, sketches, specs. etc. that are required to complete the variation, if the same to be implemented at site. This approximate estimate can be used as a guide to check the cost when the same is submitted by the Contractor.

This can be submitted to the Engineer along with a memo indicating that the site QS has taken some form of responsibility.

If the Variation is considered for action, then all documentation such as Contractor's submissions, relevant correspondence, drawings, sketches, specs, Invoices etc. to be collected, studied and finalized as early as possible.

**Status of Variation Reports**

**The Engineer's Monthly Reports to include a Section for Status of Variations, which among others to indicate the following:**

a) Total Nos. of Variation issued to date.
b) The approximate cost of Variations.
c) The impact on the Project Final value, such as within budget & time additional funds are required etc. This to be reflected in the Monthly Payment Certificate to indicate The Revised Contract Price.
d) Variations finalized to date.

Processing of Variations

All Variations to be processed at the earliest. Once a Variation cost is agreed the same to be forwarded for formal approval. This to be continuous exercise.

This will enable for early close out of Project Final Accounts as substantial portions (say around 90%) can be completed within the Contract period.

If there are difficulties in agreeing variation Rates/amounts, the disputes to be recorded and Commercial meetings to be arranged to finalize same.

A File to be maintained at site for Variations.

<u>Basic procedures that are to be followed by the Engineer while checking the Pricing of the new Rates.</u>

Contract       :
Subject        : Clause 52.1 Valuation of Variations.
                 Star Rates—Engineer's Recommendation.

Description of work
New item ref   :     *CESMM (3) Ref. if applicable
Reasons for new Rate      :
Reference of relevant correspondence, drawings, sketches etc.:
Rate proposed by the Contractor    :
Rate Recommended by the Engineer   :
Detailed computations     :
Financial/time impact to the Contract.    :

The Engineer's notes, Comments and assumptions.

## PROCEDURE TO PROCESS VARIATIONS BY CONTRACTOR

### General

I give below two cases of pricing variations to give an idea as to how what appear on the surface to be a simple exercise when used the real Quantity Surveyors skill and care become magnified and really gives an explanation as to why you find such a big gap in the salary levels of Quantity Surveyors/Cost Engineers in the construction industry. Therefore, there is a necessity to review all pricing particularly ones which involves big amounts by senior level managers such as the Contract Administrator before submission to the Employer. It is always better for the Contractor to process as and when it comes rather than wait for the Statement at Completion as approvals may be required and early processing of variation shall allow for the inclusion of same in the Interim payments to maintain healthy cash flow.

## Case One

In a new Park project, among other works, there is a work element namely—Supply and installation of **20 nos.** of light poles, complete with cabling, DB etc. for the complete operation of the system.

The BOQ amount for this item of work in the Contract is Dhs. 32,890/=. The Engineer instructs the Contractor during the later stages of the project for **another additional 5 nos.** of similar light poles to make the revised quantity 25nos. The total length between the first and the last pole remains the same as per Contract drawings.

During the detailed analysis of this additional works, the following were noted as extra or particular to this project.

- a) The Electrical capacity has increased by 25%, thus requiring additional design input and requirements of upgrading of Electrical items and equipment and possibility of redundancy of already purchased electrical materials. Some off—the self new equipment requiring delivery periods may be needed

b) As this Instruction is given in the later stages of Contract, abortive work (work already done as per programme and now requires removal and re-done), Extensions of time and Associated costs or acceleration costs in lieu to be considered.

c) Possibility of Partial completion without this work to be considered for early release of Retention or Financial charges for same to be considered.

d) Some items are not considered in the Options in order to maintain good working relationships with the Employer and the Engineer and given as notes at the end of options.

**This is an exercise to price the Variation. The options given below are not necessarily present, but given as indication to explain the worst case scenario (as it should be the case for any pricing of variations/additional works) for this type of a situation. It is also noted that there is no requirement for the amount to increase with the number of options**

The basic elements and their all in Rates for the above item of work are assumed as follows:

| | | |
|---|---|---|
| Lighting poles complete | : 20 nos. 20/300 | : 6,000.00 |
| Cabling for the light poles | : 500m approx. 4cx 16 + 1c 10 | : 10,000.00 |
| Cable Terminations to the lighting poles. 20 nos. | | : 1,000.00 |
| SMDB for Lighting | | : 5,000.00 |
| Connection from Main DB to this SMDB 100 m 4cx50mm | | : 4,000.00 |
| | Total | 26,000.00 |
| OH 15 %—say | | 3,900.00 |
| | Sub-total | 29,900.00 |
| Profit 10%—say | | 2,990.00 |
| **Total price in the BOQ** | | **32,890.00** |

## Option One:

### Only addition of poles

| | | | |
|---|---|---|---|
| Lighting poles complete | : 25 nos. 25/300 | : | **_7,500.00_** |
| Cabling for the light poles | : 500m approx. 4cx 16 + 1c 10 | : | 10,000.00 |
| Cable Terminations to the lighting poles. 25 nos. | | : | **_1,250.00_** |
| SMDB for Lighting | | : | 5,000.00 |
| Connection from Main DB to this SMDB 100 m 4cx50mm | | : | <u>4,000.00</u> |
| | Total | | 27,750.00 |
| | OH 15 %—say | | <u>4,162.50</u> |
| | Sub-total | | 31,912.50 |
| | Profit 10%—say | | <u>3,191.25</u> |
| **Total price for Option 1** | | | **35,103.75** |

## Option Two:

## Addition of poles, change of feeder cable

| | | | |
|---|---|---|---|
| Lighting poles complete | : 25 nos. 25/300 | : | **7,500.00** |
| Cabling for the light poles | : 500m approx. 4cx 25 + 1c 10 | : | ***12,500.00*** |
| Cable Terminations to the lighting poles. 25 nos. | | : | ***1,750.00*** |
| SMDB for Lighting | | : | 5,000.00 |
| Connection from Main DB to this SMDB 100 m 4cx50mm | | : | <u>4,000.00</u> |

|  |  |  |
|---|---|---|
| | Total | 30,750.00 |
| OH 15 %—say | | <u>4,612.50</u> |
| | Sub-total | 35,362.50 |
| Profit 10%—say | | <u>3,536.25</u> |
| **Total price for Option 2** | | **38,898.75** |

## Option Three:

### Addition of poles, change of feeder cable & SMDB

| | | | |
|---|---|---|---|
| Lighting poles complete | : 25 nos. 25/300 | : | ***7,500.00*** |
| Cabling for the light poles | : 500m approx. 4cx 25 + 1c 10 | : | ***12,500.00*** |
| Cable Terminations to the lighting poles. 25 nos. | | : | ***1,750.00*** |
| SMDB for Lighting | | : | ***8,000.00*** |
| Connection from Main DB to this SMDB 100 m 4cx50mm | | : | <u>4,000.00</u> |
| | Total | | 33,750.00 |
| | OH 15 %—say | | <u>5,062.50</u> |
| | Sub-total | | 38,812.50 |
| | Profit 10%—say | | <u>3,881.25</u> |
| | **Total price for Option 3** | | **42,693.75** |

## Option Four:

**Addition of poles, change of feeder cable & SMDB, change in cable for connection from main DB to SMDB**

| | | | |
|---|---|---|---|
| Lighting poles complete | : 25 nos. 25/300 | : | ***7,500.00*** |
| Cabling for the light poles | : 500m approx. 4cx 25 + 1c 10 | : | ***12,500.00*** |
| Cable Terminations to the lighting poles. 25 nos. | | : | ***1,750.00*** |
| SMDB for Lighting | | : | ***8,000.00*** |
| Connection from Main DB to this SMDB 100 m 4cx70m | | : | ***<u>6,500.00</u>*** |
| | Total | | 36,250.00 |
| | OH 15 %—say | | <u>5,437.50</u> |
| | Sub-total | | 41,697.50 |
| | Profit 10%—say | | <u>4,168.75</u> |
| | **Total price for Option 4** | | **45,866.25** |

## Option Five:

**Addition of poles, change of feeder cable & SMDB, change in cable for connection from main DB to SMDB, upgrading of MDB**

| | | |
|---|---|---|
| Lighting poles complete | : 25 nos. 25/300 | : *7,500.00* |
| Cabling for the light poles | : 500m approx. 4cx 25 + 1c 10 | : *12,500.00* |
| Cable Terminations to the lighting poles. 25 nos. | | : *1,750.00* |
| SMDB for Lighting | | : *8,000.00* |
| Connection from Main DB to this SMDB 100 m 4cx70m | | : *6,500.00* |
| Upgrading of Main DB | | *4,000.00* |
| | Total | 40,250.00 |
| | OH 15 %—say | 6,037.50 |
| | Sub-total | 46,287.50 |
| | Profit 10%—say | 4,628.75 |
| | **Total price for Option 5** | **50,916.25** |

**Option Six:**

**Addition of poles, change of feeder cable & SMDB, change in cable for connection from main DB to SMDB, upgrading of MDB, abortive Work.**

| | | | |
|---|---|---|---|
| Lighting poles complete | : 25 nos. 25/300 | : | ***7,500.00*** |
| Cabling for the light poles | : 500m approx. 4cx 25 + 1c 10 | : | ***12,500.00*** |
| Cable Terminations to the lighting poles. 25 nos. | | : | ***1,750.00*** |
| SMDB for Lighting | | : | ***8,000.00*** |
| Connection from Main DB to this SMDB 100 m 4cx70mm | | : | ***6,500.00*** |
| Upgrading of Main DB | | | <u>***4,000.00***</u> |
| | Total | | 40,250.00 |
| | OH 15 %—say | | <u>6,037.50</u> |
| | Sub-total | | 46,287.50 |
| | Profit 10%—say | | <u>4,628.75</u> |
| Total Abortive work as detailed | | | <u>***4,500.00***</u> |
| | **Total price for Option 6** | | **55,416.25** |

**Option Seven:**

**Addition of poles, change of feeder cable & SMDB, change in cable for connection from main DB to SMDB, upgrading of MDB, Abortive work, E.O.T & Associated costs for 21 days (SMDB & MDB are assumed to be off the self items)**

| | | | |
|---|---|---|---|
| Lighting poles complete | : 25 nos. 25/300 | : | ***7,500.00*** |
| Cabling for the light poles | : 500m approx. 4cx 25 + 1c 1 | : | ***12,500.00*** |
| Cable Terminations to the lighting poles. 25 nos. | | : | ***1,750.00*** |
| SMDB for Lighting | | : | ***6,500.00*** |
| Upgrading of Main DB | | | ***4,000.00*** |
| | Total | | 40,250.00 |
| | OH 15 %—say | | 6,037.50 |
| | Sub-total | | 46,287.50 |
| | Profit 10%—say | | 4,628.75 |
| Total Abortive work as detailed | | | ***4,500.00*** |
| Associated costs for 21 days as detailed | | | ***40,005.00*** |
| **Total price for Option 7** | | | 95,421.25 |

## Option Eight:

Addition of poles, change of feeder cable& SMDB, change in cable for connection from main DB to SMDB, upgrading of MDB, Abortive work, Acceleration of activities A, B, C, to make up 21 Extension costs for 21 days (SMDB & MDB are assumed to be off the self-items)

| | | | |
|---|---|---|---|
| Lighting poles complete | : 25 nos. 25/300 | : | **7,500.00** |
| Cabling for the light poles | : 500m approx. 4cx 25 + 1c 10 | : | **12,500.00** |
| Cable Terminations to the lighting poles. 25 nos. | | : | **1,750.00** |
| SMDB for Lighting | | : | **8,000.00** |
| Connection from Main DB to this SMDB 100 m 4cx70mm | | : | **6,500.00** |
| Upgrading of Main DB | | | **4,000.00** |
| | Total | | 40,250.00 |
| | OH 15 %—say | | 6,037.50 |
| | Sub-total | | 46,287.50 |
| | Profit 10%—say | | 4,628.75 |
| Total Abortive work as detailed | | | **4,500.00** |
| Acceleration costs for 21 days (Activities A, B, C, as detailed) | | | **30,005.00** |
| **Total price for Option 8** | | | **85,421.25** |

## Option Nine:

**Addition of poles, change of feeder cable & SMDB, change in cable for connection from main DB to SMDB, upgrading of MDB, Abortive work, Acceleration of activities D, E, C, to make up 21 Extension costs for 21 days (SMDB & MDB are assumed to be off the self-items)**

| | | | |
|---|---|---|---|
| Lighting poles complete | : 25 nos. 25/300 | : | **7,500.00** |
| Cabling for the light poles | : 500m approx. 4cx 25 + 1c 10 | : | **12,500.00** |
| Cable Terminations to the lighting poles. 25 nos. | | : | **1,750.00** |
| SMDB for Lighting | | : | **8,000.00** |
| Connection from Main DB to this SMDB 100 m 4cx70mm | | : | **6,500.00** |
| Upgrading of Main DB | | | **4,000.00** |
| | Total | | 40,250.00 |
| | OH 15 %—say | | 6,037.50 |
| | Sub-total | | 46,287.50 |
| | Profit 10%—say | | 4,628.75 |
| Total Abortive work as detailed | | | **4,500.00** |
| Acceleration costs for 21 days (Activities D, E, C, as detailed) | | | **35,005.00** |
| **Total price for Option 9** | | | **90,421.25** |

**Option Ten**:

**Addition of poles, change of feeder cable & SMDB, change in cable for connection from main DB to SMDB, upgrading of MDB, Abortive work, E.O.T & Associated costs for 33 days (SMDB & MDB require 15 day delivery period)**

| | | | |
|---|---|---|---|
| Lighting poles complete | : 25 nos. 25/300 | : | *7,500.00* |
| Cabling for the light poles | : 500m approx. 4cx 25 + 1c 10 | : | *12,500.00* |
| Cable Terminations to the lighting poles. 25 nos. | | : | *1,750.00* |
| SMDB for Lighting | | : | *8,000.00* |
| Connection from Main DB to this SMDB 100 m 4cx70mm | | : | *6,500.00* |
| Upgrading of Main DB | | | *4,000.00* |
| | Total | | 40,250.00 |
| | OH 15 %—say | | 6,037.50 |
| | Sub-total | | 46,287.50 |
| | Profit 10%—say | | 4,628.75 |
| | | | 50,916.25 |
| Total Abortive work as detailed | | | *4,500.00* |
| Associated costs for 33 days as detailed | | | **62,865.00** |
| **Total price for Option 10** | | | **118,281.25** |

**Option Eleven**:

**Addition of poles, change of feeder cable & SMDB, change in cable for connection from main DB to SMDB, upgrading of MDB, Abortive work, E.O.T & Associated costs for 33 days (SMDB & MDB require 15 day delivery period & air freight)**

| | | | |
|---|---|---|---:|
| Lighting poles complete | : 25 nos. 25/300 | : | *7,500.00* |
| Cabling for the light poles | : 500m approx. 4cx 25 + 1c 10 | : | *12,500.00* |
| Cable Terminations to the lighting poles. 25 nos. | | : | *1,750.00* |
| SMDB for Lighting | | : | *8,000.00* |
| Connection from Main DB to this SMDB 100 m 4cx70mm | | : | *6,500.00* |
| Upgrading of Main DB | | | *4,000.00* |
| Air Freight and clearing charges | | | *4,500.00* |
| | Total | | 44,750.00 |
| OH 15 %—say | | | 6,712.50 |
| | Sub-total | | 51,462.50 |
| Profit 10%—say | | | 5,146.25 |
| | | | 56,608.75 |
| Total Abortive work as detailed | | | *4,500.00* |
| Associated costs for 33 days as detailed | | | *62,865.00* |
| **Total price for Option 11** | | | 123,973.75 |

## Option Twelve:

Addition of poles, change of feeder cable& SMDB, change in cable for connection from main DB to SMDB, upgrading of MDB, Abortive work, E.O.T & Associated costs for 33 days (SMDB & MDB require 15 day delivery period & air freight), Financing Charges for late receipt of Retention.

| | | | |
|---|---|---|---|
| Lighting poles complete | : 25 nos. 25/300 | : | **7,500.00** |
| Cabling for the light poles | : 500m approx. 4cx 25 + 1c 10 | : | **12,500.00** |
| Cable Terminations to the lighting poles. 25 nos. | | : | **1,750.00** |
| SMDB for Lighting | | : | **8,000.00** |
| Connection from Main DB to this SMDB 100 m 4cx70mm | | : | **6,500.00** |
| Upgrading of Main DB | | | ***4,000.00*** |
| Air Freight and clearing charges | | | ***4,500.00*** |
| | Total | | 44,750.00 |
| OH 15 %—say | | | 6,712.50 |
| | Sub-total | | 51,462.50 |
| Profit 10%—say | | | 5,146.25 |
| | | | 56,608.75 |
| Total Abortive work as detailed | | | ***4,500.00*** |
| Associated costs for 33 days as detailed | | | ***62,865.00*** |
| Finance charges for the late receipt of Retention as detailed | | | **10,000.00** |
| **Total price for Option 12** | | | 133,973.75 |

**Summary of Options and the amounts.**

1	35,104/=

2	38,899/=

3	42,694/=

4	45,867/=

5	50,917/=

6	55,417/=

7	95,422/=

8	85,422/=

9	90,422/=

10	118,282/=

11	123,974/=

12	133,975/=

**Notes:**

**In addition to the above, further options can be considered with the inclusion of the following items.**

a) Now redundant materials (cables, distribution board etc.)
b) Additional Electrical Installation for the main power supply.
c) Changes in the site works for cables

d) **Claim for the loss of profit**
e) **Additional design input**

**The complete documentation such as**

**Original tender drawings, "As Built Drawings" Shop Drawings, contemporary records, proof of bank charges for interest etc. to be submitted along with pricing of the selected option.**

## Case Two

A Design & Build project to the value of US $ 200 million consisted the following as Tender and Contract documents.

1) Tender drawings giving outline scope of the work
2) Bill of Quantities to form a uniform basis for tender. The Quantities are not subject to Re-measure. However, in the event of additional works/variations, the BOQ Rates were to be used.
3) General & Particular specifications.
4) Questions & Answers during the tender period, which is to take precedent in the event of any disputes with other Contract documentation.

The design requirement for stirrups was 8 mm dia. bars which was changed to 12 mm dia. by an Engineer's instruction. This is a variation. The Engineer found a unit Rate in the BOQ which reads as SR. 2.50/kg for varying diameters and suggested to use this rate for pricing of the variation.

I, as the Contractor Quantity Surveyor objected to this principle stating that the B.O.Q unit Rate is not applicable for the following reasons.

A unit Rate consists of four basic components namely material, labour, plant overhead & profit. Material price for smaller diameters are usually marginally higher but can be ignored for this exercise. Overhead &profit remain constant. The project has re-bars of dia. 32, 25,16,12,10 & 8 mm and there is a big difference in labour & plant component for smaller diameters. Therefore a new Rate has to be used for this variation as this involves smaller dia. The concept was agreed and I was asked to submit a new unit Rate for 12 mm dia for the Engineer's review.

I calculated the tender requirement of Re-bar of 32 mm, 25,16,12,10 & 8 mm along with their unit Rates and made the substation of 2.50/kg as the average Rate in the tender. The Unit Rate for 12 mm in this exercise was 4.12/kg. The Engineer accepted the unit Rate and the same was used for pricing this variation. The final difference ( 4.12-2.50/kg) was in excess of SR. 100,000/=

**Notes:**

**The following important issues have to be taken care of while monitoring and Pricing of Additional works/variations.**

**Cumulative amount of Additional works/Variations**

The effect of the cumulative amount of all variations and additions in the Final Contract price in relation to the Original Contract price has to be monitored. Most, if not all of the Conditions of Contract ( Clause 52.3 of FIDIC) do have a provision for the Employer to recover from the Contractor the additional overheads beyond a certain percentage.(15% in case of FIDIC). The Rationale behind the Clause is that with a given set of resources used by the Contractor, the Overhead recovery to be a minimum of 85%. If the Contractor is not recovering 85%, say owing to reduction in scope, then the Employer has to compensate the Contractor for the short fall. Similarly if the Contractor recovers more than 115% of overheads, then the Employer has to be compensated as it is considered to be unjust profit. When there are excess additional works/variations, there are always possibilities for the Contractor to increase the resources and in such an event, this Clause may not be applicable. The Contractors should be aware of this and give proper notices and substantiations in such cases.

**Discounts in the Tender**

The Contractor may have been in a bad position and in order keep the company afloat particularly during tough economic times do give substantial discounts. In the event of additional works there is a possibility to use tender Rate and in that cases the discounts shall generally apply There may be completely new works, not anticipated at the time of tender and the Employers shall continue to apply the tender discounts. The Contractors should be aware of this and take necessary actions such as notices etc. to inform the Employer about the discount position on the additional works.

**Additional works/Variations to be reviewed by Senior Manager before submission.** The reason is as follows:

During my tenure as a Consultant Quantity Surveyor in one of my projects, while going through the Pricing Preamble I noted a very interesting thing namely any void less than one m2 to be ignored. This was puzzling to me as the usual standard is 0.10 m2. The project included among others various housing units and lot of Form work for concrete haunching. This had a significant effect as there are virtually no voids in the bathrooms and the Formwork quantities were higher.

While checking the Contractor's measurement submissions, I noted that the Contractor has followed the usual standard of 0.1 m2 for the voids. I do not want to take the advantage of the Contractor's errors as it is not fair and returned same and telling him to study the Pricing Preamble. This was very much appreciated by the Contractor's QS as the amount claimed was around Dhs. 200,000 less. This implies that all such preparations have to be reviewed.

**Statement at Completion**

Normally most of the Conditions of Contract do not allow any item of work that was not included in the Statement at Completion to be included in the Final Statement, other than work done during the Defects liability period.

# 8

# CONSTRUCTION FRAUDS

Frauds are part of human nature, no civilization or culture has escaped this. Construction frauds are therefore a natural phenomena. As economic situation is becoming bad the frauds are increasing. Construction is estimated to be worth up to $ 100 Billion per year in the United Kingdom alone. Some leading Construction Auditing companies are of the view that Frauds in the industry is up to 10% and the same is confirmed by the National Audit office in the United Kingdom in their report "Modernizing Construction" in 2001. Needless to say that the amounts are huge. In the United Kingdom alone the amount is approximately US $ 10 Billion per annum. All are affected by the construction frauds. By and large the Employers—be it the Governments or Individual Employers are badly affected.

It is rather unfortunate that Construction Professional Institutions, whose primary concern should be to give the value for money or the best possible value for the money spent have not come out openly with solutions for this kind of situation. My personal opinion is that all major construction projects to include a condition for the provision of Technical Auditing by competent approved Technical Auditors as traditional auditing by the Accountants is some what outdated in the complex construction environment. It is also my opinion based on the attached case studies that Employers in the construction industry should spend at least **One dollar for every Thousand dollars** spent to see whether the money is spent properly in respect of best possible value for money.

## The underlying causes for the increase in Construction frauds

Economic pressure

Cost based Contracting approaches, with the absence of incentives to the Contractor to give a competitive price.

Contracts where the scope is not fully defined until after the price is set, leading to additional works/variation during implementation.

Construction joint ventures

Deteriasion of moral & cultural values in the Society as a whole and some professionals in particular, leading to collution, conflicts of interests.

Non-proper establishments of Auditing Mechanisms

No Standardized methods or codes of construction Auditing

Non-competence of the Consultant/Employer staff which allows the Contractor to outsmart them.

It is comparatively easy to get away as prosecution of same is much more complicated.

## Frauds can be committed by

Contractor employees

Owner employees

Contractors, sub—Contractors and Consultants

Participants of successful and non—successful projects.

**Frauds can be classified as follows:**

1) False application for payment
2) Billing for work not performed.
3) Substituting or removing materials from site
4) Diverting purchases and theft of equipment and tools
5) False representations, such as using undocumented workers
6) Diverting lump—sum charges to time and material cost, which creates an idle or under production of labour and plants, where the Employer loses.
7) Manipulation of Provisional amounts or contingency items.
8) Non-payment of sub contractors and material suppliers
9) Sub Contractor collusion or conflicts of interest, which often seen in bid rigging, price Fixing, phantom hires etc.
10) Change order manipulations, giving much higher values for additional works and much less for items that are removed from scope.

**It is always better to prevent rather than cure to be the best form of action. Contracts without some form of contractor competition in order to arrive at the tender price should not be preferred. Cost plus contracts, Contracts without proper and clear scope are to be avoided as these are the playground for fraud personals. It is also evident that Frauds 1-6 given above have little or no place in an environment, where competitive tender with a Clear scope prevail. I am, as a construction Cost Professional in general and an Employer representative in particular, has a professional and moral duty to give "Value for money". Contracts without clear scope, proper tender documents and awarded without competition among pre-qualified contractors are in conflict with the above principle. Therefore, I am constrained not to deal with such frauds. I will deal with other Frauds.**

**7. Manipulation of Provisional amounts or contingency items.**

Provisional Sums have no place in a competitive tender as Provisional sums are the prescription of tender documents. Zero Provisional sums are the best tender documents. Contingency sums which are very small and are there for any unforeseen requirements, which may or may not be used.

**8. Non-payment of sub contractors and material suppliers**

In a competitive tender the responsibility is totally under the main Contractor and the Employer has no Contractual relationships with Sub-Contractors or suppliers and at the end of the day, upon Substantial completion the values of these shall be zero.

**9. Sub Contractor collusion or conflicts of interest, which often seen in bid rigging, price fixing, phantom hires etc.**

**10. Change order manipulations, giving much higher values for additional works and much less for items that are removed from scope. Outsmarting of the Contractor.**

Please refer my case studies regarding **9 & 10** above. The case studies were taken from small and medium size projects. These projects were designed and supervised by Engineering Consultants. The Tender documents contain among others

Tender and Contract Conditions

Specifications, Bill of Quantities with approx. quantities to form a uniform basis for tender. These Quantities are subject to re-measure upon completion, and were awarded after competitive tender among pre-qualified Contractors and with the Tender Analysis Report with Recommendations from the Consultant.

The Employer makes a separate payment for Design and supervision to the Engineering Consultants.

I would like to mention that in my dealings with the Contractors' entitlements, I never exercised the clout as an Employer representative and always maintained high moral values and an attitude that if and when the Contractor is entitled, then he is entitled. Unjust profits are always a concern to me.

The Case studies are given to illustrate the amounts the Employer loses owing to the above reasons. These cannot be called as frauds as per the definition. It is difficult to pinpoint the reasons or the personals responsible for such occurrence. I will leave for the reader to form his or her own conclusions.

The readers, if interested in construction frauds, can refer, what was speculated by the news media as the construction of Fraud of the centenary and which was the subject of many books—Colombo Hilton Hotel project.

**Case Study One—Over Certification by the Consultant**

Scope of work

A minor Drainage project with the Contract value of US $ 5 million.

The Protection slab over Drainage pipes was given as a slab of width 300 mm with a thickness of 50 mm. An Item for this was allowed in the BOQ (2/32/A) and the same was priced as Dhs.30.00/m.

During the execution of the Project, the Engineer changed the Protection slab as per Drainage Department standard details. In essence, the Protection slab was made to a width of 1550mm 150 mm thick, reinforced with 10 mm dia. Re-bar at 200 mm c/c in both directions.

The Contractor submitted a new Rate of Dhs.465.00/m based on Pro Rata from BOQ Rate. The Contractor submitted his analysis of new Rate as follows:

Approved BOQ (2/32/A) Rate for

300mm wide X 50 mm thick pre-cast concrete slab laid over drainage pipe was Ds. 30.00/m

Volume of Concrete for 1.00 m length as per BOQ requirements

$$300 \text{ mm} \times 50 \text{ mm} = 1.00 \times 0.30 \times 0.050$$
$$= 0.015 \text{ m}3$$

Volume of Concrete for 1.00 m length as per new requirements:

$$1550 \text{mm} \times 150 \text{ thick} = 1.00 \times 1.550 \times 0.15 = 0.2325 \text{ m}3$$

Rate for the new concrete paving slab on Pro Rata basis is

$$30/0.015 \times 0.2325 = \text{Dhs. } 465.00/\text{m}.$$

The Engineer has forwarded to the Employer with the recommendation to accept the new Rate stating that the Contractor has not considered additional requirements such as Reinforced concrete price which is marginally higher than the tender requirement of mass Concrete; Formwork Reinforcement etc.

The Final Quantity in the Contract is 400 m.

I, as Employer Quantity Surveyor found it not acceptable and made the following comments along with a detailed analysis for a new Rate to the Engineer for his observation and comments

1. The Pro Rata cannot be applied as protection slab mentioned and priced in the BOQ is a small and probably a precast element and the final requirement is a reinforced concrete item of work. Both are not similar.
2. The Engineer's assessment that the new requirement involves additional cost elements such as reinforced concrete which is marginally higher in price, formwork, reinforcement etc . . . are correct. However, the statement cannot be accepted as the all in Rate for mass concrete (Dhs.2,000/= per m3), which is used in this case is very high, more than 400% of the prevailing market Rate. Therefore, this way of pricing cannot be accepted.
3. This is a indication that there was no detail study of this Unit Rate during the tender analysis before the award of the project.
4. Contractually BOQ rates to be used if applicable. There are possibilities for changes to protection slabs owing to site conditions, and hence there is a definite possibility for use of this rate as a base. Therefore the analysis of this Rate should have been done during the Analysis of tender before the award. This is a big error in the Contract documentation.
5. This requires pricing from the first principles using the market or reasonable Rates.

## Consider 1.00 m length of Revised Protection Slab

Additional excavation & disposal

| | | | |
|---|---|---|---|
| 2/1.00 x0.40 x 0.8 | = 0.64 m3 | @ 35.00(15+20) : | 22.40 |
| Concrete: 1.00 x1.55 x 0.15 | = 0.233 m3 | @ 300.00 : | 69.90 |
| Formwork: 2/1.00 x0.15 | = 0.30 m2 | @ 40.00 : | 12.00 |

Reinforcement: 10 mm dia at 200 mm c/c

9/1.00 + 5/1.45 = 16.25m * 0.616 = 10.01 kg @ 4.75/kg : 47.55

| | |
|---|---|
| Sub-total | 151.85 |
| Allow for expansion joints etc. 4% | 6.07 |
| Total | 157.92 |
| **Say Dhs. 158/=./m** | |

After a series of discussion, the above New Unit Rate was accepted by the Contractor and the Engineer.

If, I am permitted, I would like to make the following comments regarding the above case study.

a) This appears to be an error in the Tender documentation as the requirement of the Protection slab given in tender and actual at site were to say the least ttotally technically different.

b) The Contractor who is prequalified and experienced knowingly or otherwise used a very high unit Rate, as there is a possibility for variation as happened in this case.

c) As this unit Rate happens to be very high this should have been noted in the tender analysis before the award and either the Unit Rate should been nnegotiated to bring to areasonable level or made clear to the Contractor that this Unit Rate shall not to be used as base for Variation.

d) The Engineer has erred in this instance which would have costed the Employer an additional amount of (465-158) 400 = Dhs. 122,800/= for this item of Work.

Note: 1 US$ = 3.68 Dhs. approx.

## Case Study Two—Over certification by the Consultant

### Scope of Work

The project comprised the construction and maintenance of a Steel Structure Building consisting of 41 Cold Store buildings including access roads and other Electro Mechanical Services and valued as US $ 30 million.

Generally

The Consultant submitted new Rates were reviewed by me (Employer representative) during the Finalization of the accounts for the project.

I noted a Phenomena that lot of changes were made to Electrical Installation particularly cables and SMDBS. And all new items with lesser capacities (than anticipated at the time of tender) were given very high prices. I decided to study the case further and in detail, by reviewing the original tender documents, meetings with relevant site staff etc. The results were as follows:

1. The Electrical changes were made by Service authorities and End user requirements.
2. The Consultant has used the actual paid up invoices, ignoring the original tender Rates as these were in their opinion new items as per site requirements.
3. The Contractor has underpriced Electrical Installation in the tender. However, since his overall price was the lowest the project was awarded to him. The problem was further complicated by the sudden increase in the market prices of Electrical items.

Another Technical meeting was held with the Electrical Engineers to find out the exact situations and why such changes were made. The Electrical Engineer informed that lesser capacity cables were sufficient and hence the same at present were used instead of the higher capacities and the specification remains unchanged. The same goes for the SMDBs and the new ones are marginally lower in capacities to that of the tender requirements.

After a series of meetings, discussions, it was finally agreed that in line with fairness (as the Employer is not expected to pay more than the tender prices for items which were in tender and subsequently with reduced capacity ) to base the new Rates as follows:

1. The new cable Rates to be based on the original BOQ Rates and the new Rate of the revised cable cannot exceed the next higher cable in the tender (egg. The new Rate of a 4C x16mm cable cannot exceed the BOQ price of 4C x 25mm cable.)
2. Original BOQ Rates are to be used for SMDBs. as the capacities of same were marginally lower.

# APPENDIX—A

## ELECTRICAL VARIATIONS—SMDBS

| New Rate Ref. | Original BOQ ref | Original BOQ item | New item | Qty | Original BOQ Rate | New Rate Recommended by Consultants. |
|---|---|---|---|---|---|---|
| 72 | 2/10/1B | SMDB CR 1 | SMDB CR | 1 | 20,000.00 | 38,378.44 |
| 73 | 2/10/1C | SMDB CR 2 | SMDB A 10 | 1 | 20,000.00 | 33,706.89 |
| 74 | 2/10/1D | SMDB CR 3 | SMDB A 11 | 1 | 20,000.00 | 34,320.22 |
| 75 | 3/10/1D | SMDB CR 1 | SMDB CR B3 | 1 | 15,000.00 | 51,606.22 |
| 76 | 4/10/1C | SMDB CR 1 | SMDB C1 | 1 | 17,500.00 | 26,114.44 |
| 77 | 4/10/1D | SMDB CR 2 | SMDB C6 | 1 | 17,500.00 | 30,545.78 |
| 78 | 5/10/1D | SMDB—CR-1 | SMDB D5 | 1 | 6,600.00 | 25,457.33 |
| 79 | 5/10/1D | SMDB—CR-1 | SMDB D6 | 1 | 6,600.00 | 23,694.00 |
| 80 | 5/10/1E | SMDB CR 2 | SMDB D9 | 1 | 6,600.00 | 26,356.89 |
| 81 | 5/10/1F | SMDB CR 3 | SMDB D10 | 1 | 6,600.00 | 26,009.33 |
| 82 | 6/10/1B | SMDB CR 1 | SMDB E 1 | 1 | 20,000.00 | 26668.89 |
| 83 | 6/10/1C | SMDB CR 2 | SMDB E3 | 1 | 30,000.00 | 31,386.00 |
| 84 | 6/10/1D | SMDB CR 3 | SMDB CR | 1 | 30,000.00 | 51,109.78 |
| 85 | 7/10/1/B | SMDB CR 1 | SMDB F1 | 1 | 8,000.00 | 23,992.22 |
| 86 | 7/10/1/C | SMDB CR 2 | SMDB F2 | 1 | 8,000.00 | 29,476.44 |
| 87 | 7/10/1/D | SMDB CR 3 | SMDB F4 | 1 | 8,000.00 | 26,328.00 |
| 88 | 7/10/1/E | SMDB CR 4 | SMDB CR | 1 | 8,000.00 | 49,154.22 |
| 89 | 8/10/1/D | SMDB CR 1 | SMDB G1 | 1 | 12,000.00 | 24,731.11 |
| 90 | 8/10/1/E | SMDB CR 2 | SMDB CR | 1 | 12,000.00 | 33,430.22 |
| | | | Totals | | 272,400.00 | 612,466.42 |

TOTAL CHANGES : Dhs. 340,066.42

The details of Savings to the Employer on the Consultant certified amounts

| Item | Qty/units | New Rate Recommended by Consultant | New Rate agreed | Savings Dhs. |
|---|---|---|---|---|
| 4C x16mm2+ 1C x16 mm2 | 2,500 m | 75.19/m | 28.00/m | 117,975.00 |
| 4C x150mm2+ 1C x70 mm2 | 750 m | 355.17/m | 121.60/m | 175,177.50 |
| 4C x120mm2+ 1C x70 mm2 | 145 m | 297.67/m | 105.60/m | 27,850.15 |
| New Rates 55, 72-90 | Total amount recommended by Consultant 612,466.42 | | Total amount Agreed 272,400.00 | 340,066.42 |
| | Total | | | 661,069.07 |

**The total amount recommended by the Consultant for the Variations in the Final Statement is 3,796,190.22. The above savings is around 17%.**

I would like to make the following comments regarding the Case study.

1. All new Rates in the Final Statements to be reviewed by Employer after Consultant's checking
2. The reviews to focus about the under/over priced items at the tender.
3. In the under priced scenario, the Employer should not be expected to pay escalated prices in the event the changes were made to reduce the original capacity. In such a case, the maximum is the BOQ Rate.

4. In cases, where, the BOQ quantity is small, the same is overpriced in tender, such Rates should not be considered as a base for new Rates.

## Case Study Three Over Certification by the Consultant

**Scope of Work**

The project comprised the construction and maintenance of

* Sewerage and Drainage networks with pipe diameters ranging from 200 mm to 1,300 mm and an approximate total length of 41,000 m. In addition 540 Manholes and 550 house connections for both existing and future plots and the value is US $ 22 million

*Variations*

I, as Employer representative checked in detail all variations. During our checking, we noted some important aspects as follows::

1. All Variations were priced based on Original BOQ Rates. On inquiries from the Consultant revealed that there is a Preamble note (in BOQ) stating that all Variations to be priced based on the Rates in the Bills of Quantities. The usual accompanying terminology—" if Applicable" was not present.
2. Some very high Rates (In BOQ for small quantities) were used in pricing the variations.
3. The Tender Analysis Report submitted by the Consultant did not sufficiently deal with the subject of very high Rates on small quantities at tender.
4. Substantial increases in the small quantities (with high Rates) given at tender.
5. The missing of the usual wording—if applicable for the use of BOQ Rates as a base for Pricing the variations

6. Substantial Financial impact on VOS 21,22 & 23 dealing with Asphalt works, owing to the use of high value BOQ Rates. Therefore, these Variations are to be dealt separately.

Series of meetings were held regarding the finalization of the accounts of the project. The Contractor and the Consultant were of the opinion that the Contract documents to be followed, particularly Variations to be priced as per BOQ Rates, whether applicable or not. I on behalf of the Employer presented the case under the following light.

1. It is true that in any Re-measure Contracts, some Rates are high and some are low.
2. The pricing of the variation is generally based on BOQ Rates, if applicable. The phrase if applicable means both—technically of the scope of work and the cost.
3. I, explained in details quoting the original tender Rates of other tenderers, market Rates etc., the loss of "Value for money "concept which is the central theme in a competitive tender like this.
4. As a compromise, we suggested to use market Rates for VOS dealing with Asphalt (VOS 21,22&23) owing to substantial increase in quantities (from tender to Actual) and very high tender rates and to follow BOQ Rates for other Variations.

The Consultant and the Contractor were not in agreement with my suggestion and insisted the use of BOQ Rates, which were very high for dealing with Asphalt works similar to others.

I decided to study the Variations further and decided to insert the actual quantities—Statement at Completion quantities instead of the Tender quantities in the Tender Analysis Report. This brought the cat out of the bag. The Contractor was not the lowest now. This is understandable as the difference between the tenders in the report was marginal (82.0,83.5,84.0 & 84.4 million.)

However, when Variations involving Asphalt (VOS 21, 22 & 23) were included the difference was very much big and apparent.

A Commercial meeting attended by all parties—Contractor, Consultant and the Employer was arranged. I represented the Employer. I explained in details of possible loss to Employer for the non-proper Contract documentation and the use of unrealistic BOQ Rates for Asphalt works. After heated discussions, it was finally agreed by the Contractor to price VOs 21, 22 & 23 as per market Rates and the others as per BOQ Rates.

The Final Amounts were as follows:

1. **Contractor's submission** : **Dhs. 80,063,986.66**
2. **Contractor's revised submission** : **Dhs. 77,224,757.48**
3. **Consultant's revised submission** : **Dhs. 76,148,360.09**
4. **Agreed Final amount** : **Dhs. 74,992,048.82**

## NOTES ON CASE STUDY THREE

1. It is evident from the above case study that the Employer ends up paying substantially more than required owing to the improper Tender & Contract Documentation.
2. The Auditing of the Final Account resulted in Savings of more than Dhs. One million.
3. There is a necessity to incorporate the actual final quantities instead of tender Quantities to see whether the Contractor is still the lowest. This kind of exercise shall improve the preparation of Tender documentation in the future.

## Case Study Four—Over certification by the Consultant

Scope of Work

This is a Ground + Two Building with all associated Electrical & Mechanical Works, and the project value is US $ 22 million.

The project consisted among others a BOQ item for Electrical Engineering Installations. The Contractor has given a Lump sum price of Dhs. 9,010,000/= A breakdown was requested, which included an additional item namely Supply, installation, & termination of cables and priced as 2,170,000/=. Although, an additional item was not entertained during the Contract documentation, the same was accepted as the amount was in the original tendered amount. Not accepting same, may result in the Contractor asking for variation for cabling as a missing item. The same breakdown was used for interim valuations.

During the execution, the cables were changed drastically and 4 c-240 mm cables were replaced with 4c-300 cables. The Contractor submitted a Variation amounting to **1,550,146/=** for additional cabling.

During the commercial meeting, the Contractor pointed out that the entire cabling is a missing item and they have already wrongly absorbed 2,170,000. 'this was not accepted as the Electrical equipment description included among others the necessary terminations, which implies including cabling. The Consultant was of the view that an addition/omission exercise similar to other BOQ items to be followed in the Contractor's breakdown.

Our detailed study on the Contractor's build up indicated very high material prices (around 300%) comparative low material component (56%) high labour component (25%) etc. It further indicated that their original build up is three times more than the usual.

The Analysis gave interesting reading as follows:

1. An addition/omission exercise for all cables on Contractor's submitted breakdown Rates along with Contractor's Price Analysis (Mat.56, Plant 5, Lab.25, OH & Profit 14%) gave **1,550,146/= additional**
2. An addition/omission exercise for all cables on Contractor's submitted breakdown Rates along with Revised Price Analysis (Mat. 69, Plant 2, Lab.15, OH & Profit 14%) gave **1,467,573/= additional**
3. An addition/Omission exercise based on market Rates for all cables with revised Price Analysis (Mat.69, Plant 2, Lab 15, OH & Profit 14%) gave **507,623/= additional**. This Gave another interesting reading namely the value of Original cabling was 700,732/= instead of 2,170,000 given in the Contractor's breakdown.

I, as Employer representative suggested that an addition/omission exercise based on market Rates and with reasonable Price Analysis to be followed. This we repeated is a fair approach as if some items are overpriced then there is always a possibility that some items are under priced as the Contractor's Final tender value is the lowest.

The Contractor refused to accept and insisted on addition/omission based on his breakdown with a realistic Price Analysis which amounted to Dhs. 1,467,573/= additional as above.

After further analysis, we produced a counter argument which essentially meant

1) All new cables, which were not in tender, not expected at the time of tender, therefore to be based on market Rates.
2) Actual quantity less than in tender—Contractor's breakdown Rates to be used.

3) Actual quantity more than tender—Contractor's breakdown Rates upto tender Quantity and the balance as additional not expected at the time of tender to be based on market Rates.
4) Reasonable Price Analysis Schedule.

and this exercise resulted in **a Savings of Dhs. 22,960/=**

The Contractor finally agreed to our previous suggestion and accepted the amount of Dhs.507, 623/= additional.

Recommendation

1. All new Rates in the Final Statements to be reviewed by Employer after Consultant's checking
2. The reviews to focus about the under/over priced items at the tender.
3. The Tender Documents to be prepared by the Consultant in consultation with Govt. Authorities, end user etc. to avoid additional/variations during implementation stage.
4. The possibility of Variations or changes to Design do give openings for the Tenderer to over price items where there are possibilities of variations.

**Case Study Five—Manipulation of Existing/Final Levels.**

No specific case study is given under this. However, it is important to note that there are possibilities for manipulation of levels in earth work Contracts. It is difficult to investigate these frauds after the event is completed. For example in a Road project, the Employer can lose a fortune by the manipulation of existing ground level by say an average of 200 mm. There is a necessity to check all these levels, signed, recorded to avoid such manipulations.

**Case Study Six—Under certification for Savings**

In another project, where the change of PC Rate for Tiling for a major Housing project was the subject of very interesting discussion.

A Re-measure housing project consists of 400 housing units. Each unit has a BOQ item for 160 m2 of tile flooring with the accepted unit Rate for supply & fixing as Dhs. 18/= per m2. The PC Rate for the supply of the tiling was given as Dhs. 17/=/m2. The actual purchase price of the tile was Dhs. 10/= per m2. The Engineer has after due consultations with the Contractor recommended an amount of Dhs. 448,000.00 (as detailed below) **as the savings** owing to the change in PC Rate to the Employer.

| | |
|---|---|
| PC Rates in Contract | : 17/= per m2. |
| Actual Purchase Price | : 10/= m2. |
| Therefore the deduction in PC | : 17-10 = 7/= m2. |
| Total Quantity involved | : 400 x 160 = 64,000 m2. |

**Total savings to the Contract is 64,000 x 7 = Dhs. 448,000/=**

The Employer representative rejected the amount quoting various issues such as the original BOQ Rate is not feasible, when wastage, OH & Profit etc., were considered. The Employer representative submitted an amount of Dhs.542, 720/= as Savings. This became a dispute.

The Employer representative suggested that, with the Contract mark up for OH & Profit of 15%, purchase price of tile as 17/m2 along with the wastage of say 3% the BOQ Rate of 18/= per m2 is not feasible. Therefore, some portions of this tiling amount should have been priced elsewhere; therefore a completely new approach has to be devised to arrive at the Savings. A mathematical Formula as given below with the only variable (purchase price) along with a constant—wastage factor of 3% was used.

m (constant) = PC Rate, Y(Constant) = 100-Over head-Profit

k (Constant) = Labour + Plant, M(Variable) = Accepted Invoice price

$$\frac{(1.03 \times m + \text{labour} + \text{Plant})}{Y} \times 100 = \text{Original BOQ Rate}$$

The Adjusted Rate shall be

$$\frac{1.03 \times M + k}{Y} \times 100 = 9.52/m2.$$

Therefore the Savings to be (18—9.52 ) 64,000 = **542,720/=**

Finally the Savings of **Dhs.542,720/=** was accepted by the Contractor. This is a moral issue. If there is front loading then, the same to be considered in any adjustment with the understanding that the front loading was done with the good intention of giving the best possible value for money as in the case of Advanced Payments which is an accepted phenomena in the construction field. Advanced payment, technically speaking can also be considered as Front loading.

The possibility of Frauds under the above circumstances can be summarized as follows:

a) **Total Fraud**:

**There is a possibility for the Contractor to actually buy the required tile at the Rate of Dhs. 10/= per m2 and then bring in Invoice for 17/= per m2 with the inside help or with the production of three quotations from Phantom companies and completely escape from savings.**

b) **Semi Fraud.**

**Inflated Invoice prices between 10 & 17/= m2**

# 9

# CONSTRUCTION CLAIMS, PREPARATION AND NEGOTIATIONS

There is no internationally accepted definition for Claim in relation to Construction. It is a similar case as Justice. The practical definition of a sort can be "A legitimate request for additional compensation in Cost/time on account of change in the terms of Contract." Some even tend to suggest that a claim situation arises when there is a dispute in the entitlement. Either way Claim is a bad word or not looked favourably by the Employer. I always avoid this word and use the phrase "Contractor entitlement".

There are claims in virtually any construction project owing to changes in the market conditions such as

a) Increased cost of borrowing have made Employers anxious to cut down original Capital out lay and avoid extra costs.
b) Conditions of Contract are changed to transfer increased liability to Contractors and Sub—Contractors.
c) There is stiffer competition for work leading to reduction in profit margins.

Prevention is always better than cure, therefore every effort should be made to avoid claim or additional payment as anything more than the budget is bad in the present tight economic environment.

**Sources of Disputes leading to Claim situations.**

1) Changed Conditions:.

Differing Conditions as to what is known or interpreted at the time of submission of bid to actual site conditions

2) Additional works

Additional works that is not included in the original scope requiring additional resources.

3) Delays. The delays beyond the control of the Contractor are listed below:
   Access to site
   Equipment/materials that are to be supplied by Employer not supplied in time
   Working drawings not provided in a timely manner to suit the programme of works.
   Shop drawings, Request for approvals etc. not reviewed in a timely manner.
4) Contract time; Requirement of additional time in consideration of the above.

5) Unforeseen obstructions or inclement weather, the conditions that cannot have been anticipated by an experienced Contractor, Force majeure

**Claim Avoidance**

The following can be considered as Claim avoidance:

1) Know exactly what the Contract requires as per the signed agreement.
2) Do exactly what the Contract requires without any interference.

3) Don't do anything without proper documentation.
4) Develop a Master plan or schedules along with short-term plans. Distribute same to all parties. Monitoring and continuous updating.
5) Avoid impossible bids
6) Communicate and maintain co-ordeal relationship with all parties.
7) Always remember Time is money.

Claims or additional payment along with disputes are part and parcel of modern day construction and hence the same is dealt in this chapter.

My long years in the field of construction and particularly with Employer organizations suggest that sometimes the Contractors do the submissions of their entitlements for the sake of submissions to keep the management happy as if they are likely to get something with lot of unwanted details, beating about the bush attitudes and no morals.

Normally one has to be morally correct and should have high moral values to be successful in the construction field. The Conditions of Contracts in construction is generally favour the Employer, then the Engineer. The Contractor is never a favoured party, this may be due to historical factors as Contractors in good olden days tend to charge very high profits, cheats, This is reflected in the requirements of Performance guarantees before the start of work, insurances, Prequalification of Contractors, submission of outline programmes with the availability of resources requirements at tender. This is also supplemented by the fact that different Contractors have different working methods. There are lot of subsidiary items, which the Contractor has to do but not paid. Some of the work appears to be additional on the surface but actually can be their working methods or part of the Contractor's risk and not necessary to be paid additionally as it has already been taken care of in the tender estimate.

Contractors are warned that Legal precedence such as "Good faith", "Quantum meruit" "Laesio enormis" "Audi Alteram Partem", "exgracia" etc. have very little or no impact in the evalution of the entitlement of the Contractor. The Contractor is not a favoured party in the Conditions of Contract. Therefore they are advised to follow the Conditions of Contract to the letter and spirit in order to receive their entitlements.

Just take a simple hypothetical case to understand the above. One Employer for some reason(known only to him) didn't like the completed Interlock tiles and wanted that to be removed from site. This has two scenarios as follows:

1) As these Interlock tiles are reasonably good( for him) and can be used else where the Contractor may be prepared to pay some amount to the Employer.
2) Another Contractor will expect payment from the Employer for both removal and transport same away from site.

I would like to emphasize the importance and necessity of a Notice of a Claim for Contractor entitlement. It should be noted that in certain Conditions of Contract, the Notice position is such that, if notice is not given, then the Claim is invalid. The English language dictionary meaning for invalid in relation to Claim is nil. The Contractors also to be aware of the Time bar Clauses in the Contract such as 20.1 in FIDIC which rule out the Claim in the event of non—notice. It appears that the logic is the Contractor is in a better position as to a Claim situation and by giving notice the Employer or Engineer is made to aware a Claim situation and may consider alternative action.

## Claim management

As mentioned before, every effort has to be made to avoid a claim, but if it happens then it has to managed properly as follows:

Recognize the Claim situation early or as early as possible under the circumstances

Avoid exaggerated or magnifying the claim and hash correspondence.

Keep Engineer and Employer informed

Follow the Conditions of Contract strictly and keep record.

## Claim preparation

Any Claim preparation should have the three "A "s namely

    A—Approach the problem
    A—Analyze the problem
    A—Attack the Problem

The Claim preparation to have: C E E.S as follows:

    C: Causes
    E: Effect
    E—Entitlements
    S—Substantiation
along with
    Notices:
        Conditions of Contract Clause references
        Following the Contractual procedure as per Conditions of Contract.
        Record of information or detailed contemporary records

**Presentation:**

**Format of Claim**

There is no requirement as to the format in which a Claim must be presented but the sections that are found in a well presented claim are

    Introduction
    Contractual basis of the Claim
    The details of the claim
    Conclusion and the Final amount.
    Appendices.
    The claim to be clear so that a school boy can understand.
    Precise and straight to the point
    Well supported by Contractual Clauses
    Notice requirements
    Substantiations
    Respect for all parties
    Willing to be flexible in order to reach a solution

The preparation is always for a worst case scenario. The statement "it always rains when you don't carry the umbrella" holds good as there are possibilities for problems when the worst case is not considered.

My personal experience with the preparation of Claim or additional entitlements during my tenure with contracting environment is that, I generally pass the draft to the secretary to have a look to see whether he understands the same. If the problem of understanding is related to technical nature then I ask him/her to ignore. My position is that if the secretary has a clear understanding of the facts then certainly the Consultant as Engineer should not have any problems in understanding my submissions.

Good preparation and understanding of the case is very important when it is presented for discussion. The presentation to be brief, clear and straight to the point. Beating about the bush, pleading, sir sir attitudes should be avoided. Conditions of Contract to be quoted if and when required. The general principle is one should be able to convince that there is an entitlement in the first place before coming to the details and amount. The first hurdle is the acceptance of the entitlement. Disputes in the amount is the subject of another detailed discussion. My personal opinion is high moral ground regarding the reliability of the entitlement to be there for it to succeed

I give below Three examples of Claim submissions which included among others volumes of documents, graphs, Charts, calculations etc. were rejected without even going through as the fundamental point was not taken care of:

**Case One.**

In one project, the payment terms were 60 days upon Engineer's certification of payment certificates. The Conditions of Contract also specifies as Fault of Employer as in the event of the Employer failing to pay to the Contractor the amount due under any certificate of the Engineer within 50 days after the expiration of the time stated in Contract within which payment is to be made the Contractor shall be entitled to slow down or terminate the contract by giving notice to the Employer, with a copy to the Engineer. Such action shall take effect 14 days after the giving of the notice. The Contractor was under the impression that payment shall be made within 60 days upon Engineers certification and forgot the Contractual risk of another 50 days. Contractually the Employer can take 60 + 50 days to effect the payment. Any action by the Contractor's side can only be made if payment is not made within 110 days and not 60 days. Further an additional 14 is days also required for the Contractor to commence his slow down or termination and thereby the Employer is given another 14 days to implement the action. So in actual fact the Employer can pay within 60 + 50 + 14 = 124 days upon Engineer' certification. When this point was made clear to the Contractor, he came back and said no Claim can be submitted for delay in payment. Thus, the entire preparation of the Claim is a waste of time and money.

Arguments put forwarded by the Contractor as, how a Contractor's payments for a project of 350 days can be delayed for 124 days after Engineer's certification has no Contractual basis and hence cannot be considered.

**Case Two**

The Contractor claimed for a missing item (in a Re-measure Contract) namely concrete surround around drainage pipes. The General Specification states that all drainage pipes to have concrete surrounds. This concrete surrounds were not measured in the Bills of Quantities. The Contractor has done this item of work as per specifications and made a claim as a variation.

Upon analyzing the details, the following came to light. These drainage pipes were actually under the ground slab and as an experienced Contractor, he should know that although the specifications call for such concrete surrounds the same is not required as there is already a slab on top of the pipe and thereby the safety of the pipe is assured. Further, the Contractor has not given Notice stating that this is a missing item, specifications called for but not measured therefore, if it is done the same shall constitute a variation.

We explained to the Contractor, that, suppose if there is a notice, then the Engineer would have known that there is no necessity for such concrete surrounds. The Contractor's arguments that the Engineer should have corrected the Contractor has no value as he may have assumed that the same is part of their working methods.

Again all those preparations, details etc. are a waste of time and money.

**Case Three**

A Design & Build Park project with a Contract period of 300 days was delayed by 650 days. The Conditions of Contract requires only 60 days delay to apply the maximum penalty. The Contractor as usual submitted volumes of documents, charts, graphs, calculation etc and requested an Extension of time of 650 days and associated costs.

A casual look at the headings for E, O. T included unrealistic information such as Shop Drawing approval delay 150 days, Interlock pattern approval delay 50 days etc. I raised a simple question; by going through the entire document is it not possible for the Employer to pick up a delay of just 60 days, then the Contractor answered easily possible. Then we said that is only required to apply the maximum penalty. The Contractor took back all the documents telling us that he will look out for other possibilities such as mediation, amicable settlements etc.

**It goes without saying that these kinds of preparation and submissions are to be totally avoided in order not to waste the Employer representative's time in addition to Contractors. A good understanding of the entitlements, high moral values, and good preparation are the key for the success of the Contractor entitlements.**

In an environment where a Claim or additional payment is always a bad word for the Employer as it involves additional money, this kind of no value but big preparations with volumes of details, meetings, discussions etc. bring in bad atmosphere and can sometimes influence the rejection of other variations which are actually genuine in nature.

The Contractors are also to be aware of another development in the Delay Analysis. The Association for the Advancement of Cost Engineering, International(AACE) has published a Technical document-The Recommended Practice for Forensic Schedule Analysis (RP/FSA)

AACE is an independent industry organization and is a leading professional society for Cost Estimators, Cost Engineers, Schedulers, Project Managers, and Project Control Specialists in the United States. It has been in existence since 1956 and has members in 78 countries. It is considered to be the largest organization serving the entire cost management professionals. It is the only technical organization of its kind in North America with an active committee(the Claim & Dispute Resolution (CDR) Committee) entirely devoted to construction claims dispute resolution.

In 2003, the CDR committee launched the Recommended Practice/FSA project. The intended purpose of this project was to "provide a unifying, standard reference for the forensic application of CPM scheduling in order to alleviate, if not eliminate, the confusion among practitioners regarding terminology, definitions and techniques of forensic scheduling"

The stated desired result was to "decrease the number of unnecessary disagreements concerning technical implementation and allow the practitioners to concentrate their skill on resolving disputes over substantive issues"

The approach was also stated in the RP/FSA:

Recommended Practice/FSA will define, describe and explain the usage of various forensic scheduling techniques in current use. It is not the intent of the Recommended Practice to exclude or to endorse any technique over others. However, it will offer caveats for usage and offer examples of best current practices and implementation for each technique. The focus of the document will be on technical aspects of forensic scheduling as opposed to the legal aspects. However, relevant legal principles will be discussed to the extent that they would affect the choice of technique and their relative advantages and disadvantages.

During the four years of development by the Task Force, drafts of the Recommended Practice were distributed to many industry practitioners resulting in several revised drafts that included major structural and technical changes to the document.

The delay analysis is both technical and an Art. The Contractors to decide the suitable mechanism which is benefit to them, to prepare their Claim dealing with Extension of Time and Associated Costs.

**Preparation of Contractor entitlement for Extension of Time and Associated Costs.**

Extension of Time and Associated Costs are a regular feature in any Construction project as changes do occur during the execution of a project. The changes may be owing to Employer changes or corrective action for a faulty design or to reflect Govt. Agency requirements. Some changes appear to be minor, but a combination of many such changes does have an impact on the final completion. Most Employers tend to indicate that these changes should be done within the Contract period. This may not be possible without acceleration of certain activities. If that is the case the Acceleration costs to be included in the Variation as given under the chapter pricing of Additional works/Variations.

Before the preparation of the entitlement, the Contractor should be aware of the Terminology used and the position of the Conditions of Contract in regard to delays and associated costs. Various methods are used to finally arrive at the Extension of Time and Associated cost. The Contractors need to know, the recent Phenomena in the Construction Industry such as final position of Float, which is finally decided by recent court cases in UK as Float belongs to the project; whoever gets there first can use it. The Protocol by the Construction Law Association etc.

The Contractor to list out the delay analysis which to include

Additional works/Variations which cannot be accommodated within the Contract period

Delay owing to Engineer/Employer

Delay owing to unforeseen circumstances

Impact of delay in the programme (Original Vs Actual)

Notices

An example of action to be taken by the Contractor for delay as per FIDIC Conditions of Contract are given below:

**Contract Clauses regarding Extension of Time under FIDIC.**

| Clause no | Description | Timing, Time limits Precedent Condition |
|---|---|---|
| 6.3 | Notify that planning/ execution is likely to be delayed or disrupted unless the Engineer issues a further drgs. or Instructions within a reasonable time. | In advance of delay occurring |
| 12.2 | Notify of Physical Condition or obstruction that cannot have been foreseeable by an experienced Contractor | forthwith within 7 days of event first arising. (Condition precedent to E.O.T.) |

| | | |
|---|---|---|
| 40.3 | Notice (Suspension by Engineer) 84 days or more requesting permission to proceed Work within 28 days | If permission not given within 84 days of suspension |
| 44.2 | Notify delay for which Contractor is entitled for E.O.T | Within 14 days of event |
| 44.3 | Submit interim particulars in cases of continuing effect | Within 42 days and intervals of not more than 14 days |
| 44.5 | Consulting Engineer to determine/agree mitigation measures to overcome delay. | In all cases where Contractor has given Notice under Sub-Clause 44.2 .44.3 |
| 48.1 | Notify that works are substantially complete and have passed the required tests on completion with undertaking to complete outstanding works during the Defects Liability period (48.2 applies to sectional completion) | upon completion (taking over certificate). |
| 52.2 | Notice to value Variations under Clause 52.2 | Within 14 days after the Issue of instruction under Clause 51 by Engineer (Notice is condition precedent to value under Clause 52.1) |

| | | |
|---|---|---|
| 53.1 | Notify intensions to claim additional Payment. See Clause 44.2—Contractor to keep contemporary records | Within 7 days after the event giving rise to the Claim first arises. (All Claims for additional. payments are subject to this Clause.) |
| 53.3 | Send Interim and Final account in respect of Claims for events of continuing effect | Within 28 days of such notice to Claim or such other reasonable time agreed with the Engineer |
| 53.3 | Send Interim & Final account of Claims of continuation effect. | as agreed with the Engineer on a continuing basis and within 28 days after the end of effect |
| 60.7 | Provide written discharge confirming total of Final Statement represents Full and final settlement of all monies due to the Contractor | upon submission of Final Statement |
| 67.1 | Refer dispute for Engineers decision | In the first place prior to Arbitration. |

| | | |
|---|---|---|
| 67.1 | Notify intension to commence Arbitration in case of dissatisfaction with the Engineers decision or Engineer's failure to give a decision. | Within 70 days after receiving the Engineer decision or 154 days after the date on which Dispute was referred to Engineer for a decision None within 84 days (Note: Failure to comply with Notice time reqd. will result in Engineer decision becoming Final and binding.) |

Society of Construction Law has produced a Delay and Disruption Protocol in 2002. This is not a Contractual document but it does provide recommendations and guidance to those involved with drafting contracts. The suggested methods are

As-Planned Vs As-Built
Impacted as-planned
Time impact Analysis
Collapsed as-built.

The Extensions of Time to be assessed based on one or more of the above suggestion along with all other relevant records such as Base and Impacted Programmes, notices and follow up actions as per conditions of contract (a model for FIDIC is given above) and to be costed taking into consideration of the under mentioned positions and to be submitted to the Employer at the earliest.

**Entitlement to Extension of Time.**

Applications for EOT should be made at or close to the time of the delay event. The Contractor is only entitled to extensions of time for events for which the Employer has assumed risk and responsibility, and only where those delays impinge upon the critical path. Extensions of time should be based on reasonable prediction of the delay.

**Float as it relates to Time**

Extensions of time will only be granted to the extent that the Employer Delay is predicted to reduce the total float on the activity path to below zero. This may differ if express provisions in the contract are to the contrary, and the Protocol recommends that such express provisions are written into contracts.

**Concurrency of Delay-its effects on entitlement to extension of time**

Where the Contractor delay occurs at the same time as the Employer's Delay, this should not reduce the Contractor's entitlement to an EOT. The protocol appears to disfavor any moves towards attempts to apportion blame and or to assess such blame in terms of percentage reductions in extension of time entitlements.

**Concurrency of Delay—its effect on entitlement to compensation for prolongation.**

Where the Contractor Delay and Employer's Decay occur at the same time, the Protocol suggests that the Contractor should be entitled to additional costs where the Contractor is able to separately identify those additional costs caused by the Employer's Delay. If the additional costs would have occurred as a result of the Contractor's Delay in any event, the Contractor is not entitled to recovery of those costs.

**Mitigation of Delay and losses**

The Contractor has duty to mitigate to mitigate the effects on his work, minimizing loss and attempting to prevent further loss.

However, this duty does not extend to requiring the Contractor to add extra resources or work outside the planned working hours.

**Basis of calculation of compensation for prolongation**

The objective is to put the Contractor back into the same position as if the Employer Risk Event had not occurred. Compensation for prolongation (other than variations) is to be based upon the actual additional costs incurred by the Contractor.

**Relevance of Tender Allowances**

The Tender allowances have limited relevance in the evaluation of costs relating to prolongation and disruption, caused by breach of contract or any other cause.

**Period for the evaluation for compensation**

The evaluation of the sum due for prolongation costs is based on the period when the effect of the Employer Risk Event was felt and not to the extended period at the end of the Contract.

**Acceleration**

Where acceleration measures are undertaken, payment in accordance with the contract provision should be made. Where the contract does not provide for payment of acceleration, the Contractor and Employer should agree upon what measures of acceleration are to be undertaken and the basis of payment before any such acceleration measures are undertaken.

**Disruption**

Disruption is defined as a disturbance, hindrance or interruption to the Contractor's normal working methods which affects the Contractor's efficiency.

## Claim preparation costs

The protocol recommends that, where it can be shown that the Contractor, or Employer, has been put to additional cost as a result of "unreasonable actions or inactions" in preparing or defending claims, then these costs should be recoverable.

## The Contract delays and the Contractor Entitlement can be summarized as follows:

| Contractor Delays | Employer | Third Party |
|---|---|---|
| Lack of Proper Resources | Late site access | Acts of God |
| Lack of proper planning | Later permits | Weather |
| Under estimation of durations | Differing site conditions | Labour strikes |
| Material delivery delays | Design changes/ Additional works | Service Authorities |
| Sub contractor delays | Suspension | Acts of Government |
| Failure to follow program | Nominated sub contractor/Delay in Employer supply Material/equipment | Acts of others beyond the control of the parties |

| Types of Delay | Course of Delay | EOT | Associated Costs | Penalty/ LD |
|---|---|---|---|---|
| Inexcusable Delays | Due solely to Contractor | No | No | Yes |
| Compensable | Due solely to Employer | Yes | Yes | No |
| Non-compensable | Third Party & Force Majeure | Yes | No | No |

The courts have always accepted the common sense approach. The use of computer software may not be accepted as any significant error fed into the programme was likely to invalidate the whole analysis. The Arbitrator being an expert may accept the software.

The Contractor has to examine the issue of critical delay by first testing the Original programme for reasonableness and completeness, then submitting the factual evidence to determine where time on the project was critically lost, and then seek to identify the cause of the loss time. This actually "As planned Vs As Built methodology.

It is also to be noted that when the impact of each change, or delay, fitted into the network

on the previously charted sequences, there is a possibility that Activities that were not critical prior to the new event may become Critical, and conversely former Critical activities may develop Float. Whether the change or delay affects the Critical path must be determined on the basis of conditions existing immediately prior to its occurrence.

**Models of submission of Extension of Time and Associated costs.**

**Summary of Costs,** The list is not complete, but an indication only.

1) Preliminaries or Site establishment costs
2) Bonds & Insurances
3) General destructions
4) Head Office Overheads
5) Financing Costs
6) Cost of preparing the Claim
7) Profit.

**Pricing Preambles**

This Claim has been evaluated in seven sections, all of which are self-explanatory. Rates have been used where applicable from our original tender and Contract.

Below is a brief description of how each section of the claim is priced.

1. Preliminaries or Site Establishment Costs

    These costs are obtained from using the Contract rates for Preliminaries. Preliminaries are divided into three types namely Set up costs, Running costs and demolition or removal. Only running costs were considered.

2. Bonds & Insurances.

    These costs are based on the projected cost for providing extension to Bonds and Insurances cover based on actual costs of providing the original Bond or Insurance relative to the period concerned.

3. Disruption

    General disruption has occurred throughout the period of the extension and has been evaluated as a percentage of the total cost of productive labour incurred during the extension of time. This percentage is an estimate based on the problems associated with maintaining continuity of production throughout the extended period of the contract.

4. Head Office Overheads

    As the Contract has been extended by a period of 6 months and the Contract value has not realized a recovery of Overheads on the same pro-rata basis, there is obvious under-recovery of overheads based on the Resources employed.

5. **Financing.**

   These costs have been calculated in two parts:
   a) Financing the release of late retention
   b) Financing the extension period costs.

6. **Cost of preparing the claim.**

   The actual costs as per man hours used are given under this.

7. **Profit**

   As we have used company resources in completing this project during the Prolongation period, we are of the opinion that, it is not unreasonable to request recovery on the basis of profit margin used on the original contract.

1. Preliminaries or Site Establishment Costs

   a) <u>Supervision staff:</u>

   Project Manager          Months      @

   <u>Engineers (4 Nos.)</u>
   1.                       months      @
   2.                       months      @
   3.                       months      @
   4.                       months      @

   <u>Foremen (5 Nos.)</u>
   1.                       months      @
   2.                       months      @
   3.                       months      @

4.　　　　　　　　months　@
5.　　　　　　　　months　@

Charge Hands (3 nos.)
1.　　　　　　　　months　@
2.　　　　　　　　months　@
3.　　　　　　　　months　@

b) Plants, equipment and Scaffolding
c) Power for lighting and use in works.
d) Temporary hoardings for Office & stores
e) Local Authority charges for temporary hoardings
f) Water for the works
g) Telephone/Fax services
h) Attendance on Nominated sub-Contractors
i) Site security, watching & lighting.
j) Site safety, sanitation & welfare measures
k) Site clearance and maintenance of temporary roads.

    To Summary

2. <u>Bonds & Insurances</u>
a) Performance Bond valid   from   to

 Cost of proving Original Bond

 Cost per month

 Cost of Bond during prolongation = x

    To collection

b) Contractors All Risks Policy  From     to
   Cost of providing Insurance
   Cost per month
   Cost of Insurance during prolongation—
         To collection

c) Workmen's Compensation     From     to
      Insurance
   Total wages paid in prolongation period

   Workmen's Compensation Insurance expressed
      as a percentage 1%.

   1% of total wages during Prolongation period—
         To collection

d) Third Party                From     to
   Total value
   Third party cost = 0.001 x =
   Spread over Contract period =  per month
   Cost in Pronation period   x
         To collection

Collection

   a) Performance Bond
   b) Contractors All Risks Insurance
   c) Workmen's Compensation Insurance
   d) Third Party Insurance

      Total to Summary.

3. <u>General Destruction on Original Contract work carried out after the original Contract completion</u>

   Total Productive labour costs    from    to    as taken from Labour
   for period
   
   Cost analysis
   
   Allow 10% for general disruption, productivity deficiencies caused by restricted access and lack of continuity of work
   
   _____

To summary                                                      _____

4. <u>Head Office Overheads</u>

   The Overhead recovery tendered for was Dhs. Y    at 8% for
   
   8 month Contract period = AED.    Per month

However, the value of the Contract has increased, so the contribution to the overheads having been spread over the unit rates will have an increased pro-rata to the increased scope of work, i.e

   Therefore    X.    Final Contract Value (anticipated)
   = Overhead recovery on executed Final Account.
   Therefore    X.    (Estimate only) =

However, as the programme of work has been extended considerably, the actual monthly overhead recovery will now have decreased to

   K                = A.E.D.    Per month overhead recovery

The Overhead loss is therefore:

(Monthly tendered − actual monthly × Total Contract
   contribution           contribution)             period

                                                         ————————

i.e. (     −     ) x months   = To summary   ————————

5. **Financing**

   Financing extension of time costs.

   Prolongation period is 6 months.

   Cost of Financing 6 months at 11% per annum = 5.5%

   Financing cost = Total Cost x 5.5%

                =           To Summary.   ————————

6. **Preparation of the claim**

   Claim specialist        days @

   Cad Draughtsman     days @

   Secretary                days @

   Consumables lump sum

                         To Summary.   ————————

7. Profit

Total Value 1 to 6     =   ————————

9.5% of profit as per tender = To Summary   ————————

**Summary of Costs.**

1) Management & Engineering Staff
2) Bonds & Insurances
3) General destructions
4) Head Office Overheads
5) Financing Costs
6) Preparation of the claim
7) Profit—

        **Total**                      **KKK**

**Some Commercial factors influencing the Contractors claim.**

The overall total loss incurred on the project.

The severity and complexity of events experienced both on and off the construction site.

The parties involved and their respective responsibilities to other.

The incidence of lack of financial resources

Labour and material shortages

Requirements to pay for third parties such as sub-Contractors & suppliers.

Application of Delay penalties by the Employer.

Has the Employer recognized the problem and budgeted for same. Contractor or sub-contractors loss of opportunity to use the resources much more profitably on other projects.

Lack of claim preparation staff or financial resources to go for litigation.

One final word about claims is that History has shown that it is always better to reach a negotiated settlement owing to:

1) Wrong doers pay heavily in courts.
2) The necessary financial and manpower required to support Arbitration/litigation may be absent or lacking.
3) There may be delay in the settlement.
4) Negotiation can give better result and can be done for a fraction of the other costs.

Notes:

It should be noted that each project is unique and what is applicable for one not necessarily applicable for another, therefore the summary of costs shall be different. It should be emphasized that the Initial Programme is to be a good one that reflects the needs of the project. Time Impact monitored on a regular basis to calculate the delay and allocations of such delays to the parties responsible for same is the most practical way of calculating the Extensions of Time. The Employer and the Engineer should be kept informed all the time as per Conditions of Contract. Time bar provisions, if applicable should be carefully considered.

Cost of preparation of claim and profit are usually not considered in the evaluation of associated costs. However, The Law Society Protocol seems to agree for same and since the Contractor has the privilege of claiming anything, so he can include these.

**It is the responsibility of the Contractor Quantity Surveyor to prepare all the Contractor entitlements properly and submit to the management before submission to the Employer.**

**It is the top management's decision to pursue the Claims in full or part or not all as they may have other much bigger commercial concerns such as maintaining good relationship with the Employer for future work etc.**

# 10

# CONSTRUCTION DISPUTES, MECHANISMS FOR RESOLUTIONS

The Construction disputes were initially resolved by the normal courts. Arbitration which is binding like the courts, was introduced owing to the complexity of the problem and there is a necessity for the requirement of Technical expertise which the court system largely lacked, to resolve the issues,

Now there is a shift in the aspects of resolving the construction disputes. The parties to the disputes do not want to be mere observers and do want to be actively involved in the resolution solving process. This also reduces the overall cost of resolving disputes as outside parties such as Lawyers and Arbitrators have to be paid, whereas the existing In-house personal, who were involved in the actions can be used. So much, so that it is now an internationally accepted practice in all construction contracts. All such resolution processes other than the traditional court system are called **Alternative Dispute Resolution**, and denoted by **ADR.**

## History

The modern root of the Alternate Dispute Resolution denoted by ADR are usually traced to a meeting of an in-house counsel in the united states about 30 years ago (many of whom were from the lucky 500 companies). The group was annoyed with the costs of litigation, the time it took and the fact the parties to the disputes felt excluded from the litigation process.

The attendees agreed among themselves that if they had disputes with each other, they would attempt ADR before resorting to traditional litigation. Representatives of the companies present signed an "ADR "pledge to indicate their support for ADR over litigation in appropriate circumstances. Out of the meeting, the center for Public Resources (CPR) was formed through which the in-house counsel could access neutrals and obtain ADR process. The CPR developed a Roster of mediators and arbitrators, many of whom were retired judges and distinguished attorneys.

ADR was introduced in the 1980s for the resolution of commercial disputes. The advantages of the ADR techniques were quickly recognized by judges, following civil procedure rules in 1999, have discretion to direct the parties to enter into ADR. This is further evidenced by the introduction of ADR as mandatory in the Contracts signed for World Bank and other leading banks.

## Forms of ADR

### Mediation

This is the most common form of ADR and the service offers the opportunity for a fresh approach to negotiations and to reduce the cost of conflict in time and money. Flexibility and confidentiality are important features.

Mediation has now become the ADR process of choice. It is seen as a way to develop creative solutions, leave the decision in the hands of the clients, and maintain ongoing business relationships. The ability to create an outcome that serves the interests of the parties liberates litigants from risks, uncertainties and loss of control of the outcome that occurs when a dispute is turned over to a third party adjudicator, a judge or an arbitrator.

The skilled mediators have the tendency of "getting to yes" situations by looking at disputants underlying interests as disputes need not always be a battle of rights. They can also experiment with interest-based mediation, where a mediator who is an expert in negotiations help disputants uncover underlying interests in a search for a joint mutual gain, and look for a resolution that made more sense to both parties than proceeding to litigation.

The settlement rates are high. For example ADR chambers which operate in Canada, the UK and internationally, the settlement rate at mediation is about 90%. Satisfaction among lawyers and clients with the mediation process is very high.

Even where cases do not settle, lawyers and clients tend to agree that the process was beneficial in clarifying and streamlining issues, and understand positions and underlying interests.

There is widespread acceptance both within the legal community and beyond the lawyers and their clients must become aware of the possibilities of mediations and should understand how the process works.

With the opportunity for creative solutions for a fast and inexpensive resolution that the parties develop and buy into, and for a process that enhance rather than destroys on-going business relationships, mediation is likely to grow rapidly in popularity over the next years in the UK.

As lawyers become more sophisticated consumers of mediation services, they and their clients will pick and choose mediators best suited for a given dispute and the number and range of competent, qualified mediators will expand.

## Arbitration

Arbitration is a binding method of disputes resolution governed by statute. The appointed arbitrator considers the evidence presented by both parties and then issue an award, which is enforceable by the courts.

Arbitration has been a traditional alternative to court based litigation. It has been widely used in commercial disputes, particularly those involving international disputes and matters where technical expertise on the part of the decision maker is important. Procedures used in arbitration can vary from informal to rules that essentially mirror that of courts. The benefit of arbitration includes its confidentiality, flexibility, speed, the expertise of the arbitrator and it is usually cheaper than courts.

## Conciliation

Conciliation is an established feature of industrial relations. The Conciliator brings the parties together and act as facilitator in their discussions. The Conciliator is not a decision maker.

## Adjudication

Adjudication is a process for the early, fast summary, temporarily binding, but interim resolution of disputes. This assures the continuity and completion of the project, which now applies compulsorily in some construction contracts.

## Med-AR

Med-AR is a process in which the parties proceed through a facilitative mediation process and if the matter does not resolve completely or if certain issues remain unresolved following the mediation process, the mediator is entitled to decide the issues as an Arbitrator.

## Mini-Trial

Mini-Trial is a process which can be useful to companies in dispute in complex matters. Typically, the senior executives of the companies, with authority to settle, form a panel together with a suitable neutral as chair of the panel.

The legal representatives for both parties make a presentation of their case in a summary fashion without witnesses or with minimal number of witnesses. The presentation to include relevant evidence followed by argument. The executives thereby have the opportunity of assessing the strengths and weaknesses of their case and that of the opponent. Following the presentation, the executives negotiate with each other, with or without the assistance of the neutral, attempting to find a mutually acceptable resolution of the issues. The neutral can function in a facilitative median manner, explore the issues needs and concern of the parties, or alternatively evaluative and provide a view of the issues for consideration of the parties. This process is non-binding and without prejudice to legal rights.

# 11
## CONCLUSION

The book shall be a useful tool for young Engineers who seek careers in the Construction Industry in general and particularly for those who deal with the costing and Auditing side of construction. It can also be used for small and medium size Contractors as it deals with the functions such as Conditions of Contract, Types of Contract, Risks and the working of the Construction Industry in general. The emphasize is on the costing aspects of Construction with examples of Pricing both for tender and additional works. This also gives interesting reading with case studies under construction frauds for those who want to have careers in the Technical Auditing of construction, which is gaining ground as Employers are demanding more and more for the money spent.

# Profile of the Author

The author is a Chartered Quantity Surveyor,& Arbitrator and has been working as Quantity Surveyor for Dubai Municipality since 1999. Having been working for a premier Government Institution dealing with construction projects, handled projects of various sizes.

**The Author has a rare mixture of Technical Qualifications coupled with long years of experience in the field and in teaching in an unique position to understand both the need of the Industry and the students.**

**Email:** malebbe@dm.gov.ae.   Tel: + 04-2064558/04-2341303
almameer@hotmail.com

**PROFESSIONAL QUALIFICATIONS:**

    Professional Member of the Royal Institution of Chartered Surveyors, London **(MRICS)**
    Associate Member of the Australian Institute of Quantity Surveyors **(AAIQS)**
    Associate Member of the Chartered Institute of Arbitrators, London **(ACIArb)**
    Elected as an Associate Member by the Construction Surveyors' Institute, London in 1980 (A.M.C.S.I.)
    Elected as a Member by the Institute of Engineers & Technicians, London in 1981 (M.I.E.)
    Successfully completed the Civil Engineering Course in 1969.
    Successfully completed the Builders' Quantities II course in 1975
    Successfully completed the Technical Teachers' Training Course conducted by The University of Moratuwa, Sri Lanka in 1976.

# EXPERIENCE

**Quantity Surveyor** - 35 years

**Teaching** - 5 years.

**Countries worked** - Sri Lanka, Oman, Bahrain, Saudi Arabia, U.A.E.

**Personal contributions.**

Presentation of a research paper titled "Auditing of Construction Projects Final Accounts at the Pacific Asia Quantity Surveyors Congress.

Submission of Proposals for Technical Auditing of Construction Projects Final Accounts.

Submission of Proposals for the Introduction of Quantity Surveying in the Technical College curriculum in Oman.

Publication of Article under the title "Construction as an option to international Investors in The Building Economics(December 2011). Thus making the QSS to think that they can after all become as Financial Advisors to International Investors and earn in excess of US$ 100,000/= a week.

Involved in Teaching Quantity Surveying courses at Technical Colleges in Sri Lanka.

Wrote a book titled "How to be successful at competitive Interviews" incorporating my successful QS interview experiences. ISBN 978-1-84903-205-6. To be published by Schiel & Denver Publishing Limited, Great Britain in March 2013.

www.ingramcontent.com/pod-product-compliance
Lightning Source LLC
Chambersburg PA
CBHW022019170526
45157CB00003B/1294